INTERNATIONAL CENTRE FOR MECHANICAL SCIENCES

COURSES AND LECTURES - No. 92

EKKEHART KRÖNER
UNIVERSITY OF STUTTGART

STATISTICAL CONTINUUM MECHANICS

COURSE HELD AT THE DEPARTMENT
OF GENERAL MECHANICS
OCTOBER 1971

UDINE 1971

SPRINGER-VERLAG WIEN GMBH

Originally published by Springer-Verlag Wien-New York in 1972

ISBN 978-3-211-81129-0 ISBN 978-3-7091-2862-6 (eBook)
DOI 10.1007/978-3-7091-2862-6

PREFACE

The scientific branch of statistical continuum mechanics, as the name says, serves the purpose of treating statistical problems in continuum mechanics. The first important successes were achieved in the field of turbulence in the thierties of this century. In recent time the statistical continuum mechanics has experienced a large extension by including the complex field of materials of a heterogeneous constitution such as polycristalline aggregates, multiphase mixtures, composites etc.. It is the main aim of the statistical theory in this field to calculate the effective properties from the given properties of the constituents and from some information about their structural arrangement. Important progress has been made in the case of elastic materials. Here the names of Beran, Molyneux, Lomakin, Volkov, Klinskikh, McCoy, and Mazilu deserve particular mention.

The great fluctuations in stress and strain which accompany plastic deformation lead to the expectation that the final theory of plasticity, too, will contain statistical elements. Investigations in this field are still very preliminary, however. So it is not clear at all whether the statistical theory should be developed along the lines of approach to turbulence or whether it should be a statistics of dislocations which, as is well-known, are the fundamental sources of the stress and strain fluctuations.

In this course I give, after the introduction in chapter I, a somewhat restricted but, I hope, still sufficiently detailed outline of the mathematical theory of probability and statistics (chapter II). Today, all the concepts and notations presented here are standard, hence I have not tried to be particularly original in this part of the course. The systematics of chapter II and of chapter III in which the mathematical theory is combined with physical laws into the formulation of statistical continuum mechanics follow largely the excellent monograph of M. Beran "Statistical Continuum Theories", Interscience Publishers 1968. Throughout this course I have tried, as far as possible, to employ the notation of Beran.

This should facilitate for the reader of the present text the study of the more comprehensive presentation of Beran.

In chapter IV the field of statistical turbulence which may already be considered classical is outlined very briefly, just to accustom the reader to the statistical approach. It also provides provides an idea of the form a statistical physical theory may assume.

Chapter V is the central part of this text into which also some of my own original ideas have been embodied. In my opinion this chapter contains the most important results in the field. I mention the discovery of Beran and McCoy and of Mazilu, which shows that the macroscopic theory of linearly elastic heterogeneous materials is in some sense congruent to the non-local (linear) theory of elasticity. In this connection also the particular role played by the perfectly disordered material is pointed out. The theoretical concept of the perfectly disordered material was introduced by me when I was a research visitor at the Applied Science Division of Harvard University in 1967. I am very grateful to Professor B. Budiansky who at that time drew my attention to the just published work of Beran and Molyneux and who assisted me during the first difficult steps in this new area.

The dynamical theory of heterogeneous materials is much less developed than the statical theory. Hence I have omitted it completely and only pointed to certain additional difficulties which enter in the transition from statics to dynamics.

Chapter VI very briefly introduces a few remarks on stochastic elements which are inherent in the phenomenon of plasticity. These remarks are mainly meant to draw the attention of those interested in plasticity to a different kind of approach which could well develop into a new important branch of statistical continuum mechanics.

The field of turbulence, elasticity and plasticity are not the only fields within continuum mechanics where problems of a stochastic nature occur. I mention the area of flow through porous media which is treated by Beran. Stochastic processes certainly also take place in visco-elastic media. There does not seem to exist much work in this field. It is for the reason of compactness of this course that the present text has been confined to the theories of turbulence, and hetero-

geneous materials and a little bit of plasticity. I believe that inspite of this restriction the problems treated here form a rather representative cross section of the problems of continuum mechanics - or even of continuum physics in general - which require stat istical treatment.

It is a pleasure to thank Dr. B. K. D. Gairola for his thotough reading of the manuscript, Dr. J. Burbach, Dipl. Phys. J. Koch and Professor S. L. Koh for numerous discussions on the effective elastic moduli of heterogeneous matrials, Mrs. H. Wetzlberger for the careful and speedy typing of the manuscript and the Deutsche Forschungsgemeinschaft which has supported my own research in the field. Finally, I would like to thank all members of CISM for their great hospitality which made my stay at the Centre so pleasant.

Udine, October 1971

E. K.

Since this preface was written, a number of important new results in the field of interest have been obtained. It seems to me that the monograph will greatly gain in value if at least a brief report on such results were included. Beside a few corrections of the original text I, therefore, have inserted the section 5.20 and new references which lead the reader to the latest work in our area.

Stuttgart, January 1973

E. K.

CHAPTER I

Introduction and programme

1.1. Introduction

Reflections on the nature of probability and statistics have been made by human beeings ever since they learned to think. It is a common feature of daily life that people try to calculate their chances in certain situations. I only remind of the numerous games of hazard to which already the old cultural nations paid homage. The fact that everybody is well acquainted with such games is the reason that dice, playcards etc. are widely used in all texts on probability and statistics.

The most spectacular success of statistical theories in physics has been gained by statistical mechanics, sometimes called statistical thermodynamics. Here it was possible to derive the phenomenological laws of thermodynamics from the classical laws - later from the quantum laws - of mechanics.

Whereas the mathematical theory utilized mainly in particle mechanics is that of ordinary differential equations it is usual to formulate continuum mechanics in the form of partial differential equations. A similar difference in the mathematics employed occurs when we compare statistical particle mechanics and statistical continuum mechanics. So the fundamental formulation of statistical continuum mechanics is given in terms of probability density functionals whereas in statistical particle mechanics one uses probability density functions. Hence the mathematics needed in statistical continuum mechanics is functional analysis. However, as this branch of mathematics is not sufficiently well developed one has to use other mathematical methods.

We usually divide the science of mechanics into particle mechanics and continuum mechanics. The distinction between statistical mechanics and statistical continuum mechanics corresponds to this division. Although the general interest in particle mechanics and continuum mechanics is of the same order of magnitude, this is not the case with the corresponding statistical theories. This fact is strange if one considers that thermodynamics deals not only with discrete particle systems but also with continuous matter. Indeed, it is suggestive to assume that the thermodynamical behaviour of continuous matter can be explained by statistical continuum mechanics in a similar manner as it is done with particle systems by statistical mechanics.

The problems treated in statistical continuum mechanics are, however, of a quite different nature. So neither the statistical theory of turbulence nor the statistical theory of heterogeneous materials, i.e. none of the two theories which we mainly present in this text, have anything to do with thermal phenomena. It follows from this fact that statistics is a more general concept than heat.

Statistics as well as probability are mathematical concepts. They acquire a physical meaning only when combined with physical laws. Clearly, these laws are those of continuum mechanics when we construct statistical continuum mechanics.

In order to see how statistical problems appear in mechanics let us consider a very simple physical system, the harmonic oscillator. Its equation of motion has the well-known form

$$m \frac{d^2x}{dt^2} + kx = F \tag{1.1.1}$$

where x, m, k, F denote displacement, mass, spring constant and driving force respectively. The solutions $x(t)$ of eq. (1.1.1) depend on the form of m, k and F. If one of these quantities is a random fucntion of time then $x(t)$, too, will be random in time. It is then called a "random variable" which underlines the fact that we deal with "stochastic processes".

Let us consider the case that F is a random function of time. An important point is that in statistics one never has the full information on a random function, i.e. the information which also includes the details of the fluctuations. In many cases we are not even interested in these details of the given random function.

The problem is now to prescribe the known information in a rational way. It could be rational, for instance, to infer the information piecewise, by giving first a rather rough information, then a little bit more and successively additional details until the whole known information is provided. We shall learn later that a systematic description of the given random function by means of correlation functions is very well adapted to the above idea of information supply. The problem of the theory consists then in producing a maximum of prediction about $x(t)$ from the information given in this way about $F(t)$. Again it appears to be sensible to use a description of $x(t)$ in terms of correlation functions.

This short illustration demonstrates one of the fundamental features inherent in any statistical theory. This is the particular manner how information is processed in statistics.

Many equations in physics have a structure which is similar to that of the oscillator equation. They contain terms accounting for inertia, resistance, and driving forces. It is easy to see how statistical problems arise in all these cases. In fact, statistics is of great universality in physics, comparable, perhaps, only to wave propagation. It is certainly no accident that waves and statistics are the fundamentals of elementary physics, one representing order, the other disorder.

1.2. Programme

This section provides a short guidance through

this text. The systematic course begins with an exposition of the mathematical theory of probability and statistics in chapter II. Since textbooks in this field usually comprise some hundred pages, our representation of 25 pages necessarily is very restrictive. We mainly explain those concepts and notions which occur in the physical theories treated later. This brief summary certainly cannot replace a proper study of the mathematical theory of probability and statistics.

In chapter III we make the transition from mathematics to physics by combining the mathematical theorems with physical laws. Once more we consider the example of the harmonic oscillator, to illustrate the general method. The formulation of statistical continuum mechanics does not offer additional difficulties.

In chapter IV we pay attention to the first important and successful field of statistical continuum mechanics, the field of turbulence. Here we see how the foregoing more formal developments can be applied to actual physical problems and even obtain successful solutions. Chapter IV also prepares the ground for the chapter V on the theory of heterogeneous elastic material which I have chosen to be the central part of this text because I have been working originally in this field during the last five years. In my opinion the statistical approach has been particularly successful in this area and further progress is to be expected.

Chapter VI finally tries to explain what plasticity and statistics have in common. It is hoped that this exposition will stimulate workers in the field of plasticity to try statistical ideas towards the further development of the macroscopic theory of plasticity.

As mentioned in the preface, we have widely adopted the notation of Beran's monograph. Since this book is frequently quoted we employ an extra quotation symbol [B] for it. All references are given chapterwise at the end of the course. Beside references the list of literature also includes texts which are recommended for study by the reader. Furthermore a number of problems arranged sectionwise are provided at the end of this book.

CHAPTER II

The mathematical basis of probability theory and continuum statistics

2.3. The axiomatic basis of probability continuum statistics

A. Kolmogorov develops the theory of probability starting from a small set of definitions and axioms from which the whole mathematical framework is derived. This is the usual manner in which also other mathematical (and physical) theories are constructed. As an example the group theory may be mentioned.

Not so much for practical use, but rather in order to give an idea of the nature of this axiomatical construction of the theory the definitions and axioms are listed below.

a) Definitions

Let ξ, η, ζ ... be certain elements which we shall call elementary events and let E be the set of these elements.

Example: If we toss two dice the elementary events are the 36 possible outcomes (1,1), (1,2) ... (1,6), (2,1), (2,2) ... (2,6) ... (6,1), (6,2) ... (6,6).

Let A be a subset of E. We call A a random event or, simply, an event, and F the set of all A's.

Example: A subset of the above E is the set A_1 of the outcomes showing an even number or the set A_2 of the outcomes (1,1), (2,2) ... (6,6) etc.

In particular, the elements ξ, η, ζ ... themselves are subsets of E, hence also elements of F.

b) Axioms (*)

I. F contains within it the sum, difference, and product of all of the subsets of E (of all A's).(**)

II. F contains the set E.

III. To each set A in F is assigned a non-negative real number P(A). This number is called the probability of the event.

IV. $P(E) = 1$.

V. If A and B have no element in common, then $P(A+B) = P(A) + P(B)$. Here $A+B$ is the set which contains all elements of A and B.

This list of axioms is complete if the number of elements is finite. In the situations considered in continuum statistics an additional axiom is needed, namely,

VI. Postulate of continuity.

This axiom ensures that the probabilities are of such a nature that they can be described by probability density functionals.

c) Comments

Axiom II can easily be included into axiom I.

(*) Cited from [B], p. 18, 19

(**) Hence, F is a field (of sets). This is the terminology of Hausdorff, Mengenlehre, 1927, p. 78.

The difference $A_1 - A_2$ is defined only if A_2 belongs to the set A_1. This is, for instance, the case in the above example. The product AB is interpreted as the subset of E the elements of which are both in A and in B. In contrast to this the subset $A+B$ consists of all elements which belong either to A or to B.

The assignment of probabilities to events will form an important part of the physical consideration of the concerned problem. Of course, $P(E)=1$ means, that the probability 1 is assigned to the event E which results in an outcome of any of the elementary events of E. In fact, according to axiom V one obtains that

$$P(E) = P(\xi) + P(\eta) + P(\xi) + \ldots$$

which means that $P(E)$ is the probability that the outcome is either ξ or η or ζ or ... .

2.4. Conditional probability and indipendence

In the next sections we discuss briefly a number of further definitions and theorems which are based on the above listed elementary definitions and axioms.

a) Conditional probability

The probability of the outcome "6" in a tossing experiment with one die is 1/6. However, if we have the information that the showing of the die is "even" then the probability of the "6" is 1/3. The information "even" is the extra condition in this example. If we call B the event "6", A the event "even" then we write $P_A(B) = P_{even}(6) = 1/3$. $P_A(B)$ is read "probability of B , given A ".

$P(AB)$ is the probability that the outcome is both in A and in B . Since the event AB requires the outcome "6" the probability $P(AB)$ is 1/6 in our example. So we see that

(2.4.1) $$P_A(B) = P(AB)/P(A)$$

in the present case. It can be shown that eq.(2.4.1) is true for any other experiment involving events A and B . From an axiomatic standpoint eq. (2.4.1) is to be considered as a definition of conditional probability. This notion is extremely important in probability theory.

b) Independence of events

Two events A and B called mutually independent if

(2.4.2) $$P(AB) = P(A)P(B)$$

In this case $P_A(B) = P(B)$ (compare eqs. (2.4.1) and (2.4.2)), i.e. the information A has no influence on the probability of B: both events have nothing to do with each other.

Example: If we know in an experiment with two dice that the first die shows a "5", then this information has no consequence for the showing of the second die.

The calculation with independent events is much simpler than that with dependent ones. However, dependence is often the very essence of a physical situation.

2.5. Distribution functions and density functions

a) Probability distribution functions

Assume, our particular experiment results in outcomes which correspond to real numbers u in the range $-\infty \ldots +\infty$. u is called a random variable because we are interested in experiments of chance. For instance, u can be the x-component of the velocity of a turbulent flow at time t and position $\underline{x}$.

The probability that the outcome of the measurement lies between $-\infty$ and u is denoted by $F_1(u)$. $F_1(u)$ is called probability distribution function or simply distribution function. The 1 in F_1 indicates that the distribution is 1-dimensional. Obviously

$$\lim_{u \to -\infty} F_1(u) = 0 , \qquad \lim_{u \to +\infty} F_1(u) = 1 \tag{2.5.3}$$

and the probability that u lies between a and b is

$$P_1(a \leq u \leq b) = F_1(b) - F_1(a) . \tag{2.5.4}$$

If we are interested in three components u, v, w of the velocity at point $\underline{x}$ at time t, then we introduce the 3-dimensional distribution function $F_3(u,v,w)$ which gives the probability that the outcome of the experiment lies in the range $-\infty \ldots u$, $-\infty \ldots v$, $-\infty \ldots w$. We find that $F_3 = 0$ if u or v or w approaches $-\infty$ and that $F_3 = 1$ if u and v and w simultaneously approach $+\infty$. For $w \to +\infty$ one obtains $F_3(u,v,w) \longrightarrow F_2(u,v)$ and similarly for $v \to +\infty$ follows $F_2(u,\infty) = F_1(u)$.

One can use the same definitions in the case where u, v, w are the x-components of the velocity at three points $\underline{x}_1$, $\underline{x}_2$, $\underline{x}_3$ either at the same time t or, more general, at three different times t_1, t_2, t_3.

As a generalization of this we construct the n-dimensional distribution function $F_n(u_1, u_2 \dots u_n)$ where the u's are the outcomes of measurements at n points $\underline{x}_1, \underline{x}_2 \dots \underline{x}_n$. If the quantity of interest is a vector $\underline{v} = (u, v, w)$, say, as it is in the problem of turbulence, then a distribution function of the form

$$F_{3n}(u_1 \dots u_n ; v_1 \dots v_n ; w_1 \dots w_n) \equiv F_{3n}(\underline{v}_1 \dots \underline{v}_n)$$

can be used. In the case of tensors of second rank distribution functions F_{9n} etc. can, similarly, be formed.

b) <u>Continuous distributions, Functionals</u>

This being a course on statistical <u>continuum</u> mechanics we are especially interested in measurements over a continuum of points $\underline{x}$. This implies that the discrete outcomes u_1, $u_2 \dots u_n$ now become dense and form fields $u(\underline{x})$. In the example of turbulence we are interested in the velocity field $\underline{v}(\underline{x})$.

The generalization to be considered here leads to an ∞-dimensional distribution function which is usually written as a functional $F = F[\underline{v}(\underline{x})]$. We call F a space functional of the velocity because F depends on the values of $\underline{v}$ in space. A quantity which depends on the values of $\underline{v}$ at a fixed point but at various times t is a time functional of $\underline{v}$. The general case is, of course, the space-time functional, which is written for instance as $F = F\left[\underline{v}(\underline{x},t)\right]$.

One can speak of functionals as functions of functions. They occur in many parts of continuum and field physics.

Example: If the stress at point $\underline{x}$ in an elastic continuum depends on the strain at the same point $\underline{x}$ only, then the stress is a function of the strain. However, if the stress at $\underline{x}$ depends simultaneously on the strain at many or all points $\underline{x}'$ in the body then the stress is a space functional of the strain. If the stress at $\underline{x}$ and t depends on the values of the strain at $\underline{x}$ at some or all former times t' then the stress is a time functional of strain. The general case is, of course, that the stress is a space-time functional of strain.

The stress as a space functional appears in the so called non-local continuum mechanics. The time functional is important in materials with memory. We shall come back to the functional formulation of probability later.

c) Probability density functions

If the range $b-a$ becomes infinitesimal we write da and have from eq. (2.5.4)

$$P_1(a \leq u \leq a+da) = F_1(a+da) - F_1(a) = \frac{dF_1}{da}\,da \tag{2.5.5}$$

assuming that F_1 is continuous in u (and a). We call

$$P_1(u) = dF_1(u)/du \tag{2.5.6}$$

the probability density (also density function) belonging to u. Obviously, $P_1(u)du$ is the probability of an outcome in the range $<u \dots u+du>$. Of course,

$$(2.5.7) \qquad \int_{-\infty}^{\infty} P_1(u)\,du = 1 .$$

Higher-dimensional density functions are correspondingly formed:

$$(2.5.8) \qquad P_2(u,v) \equiv \frac{\partial^2 F_2(u,v)}{\partial u \, \partial v}$$

$$(2.5.9) \qquad P_n(u_1, u_2 \dots u_n) \equiv \frac{\partial^n F_n(u_1, u_2 \dots u_n)}{\partial u_1 \partial u_2 \dots \partial u_n}$$

$$(2.5.10) \qquad P_{3n}(u_1..u_n; v_1..v_n; w_1..w_n) \equiv \frac{\partial^n F_{3n}(u_1..u_n; v_1..v_n; w_1..w_n)}{\partial u_1..\partial u_n \, \partial v_1..\partial v_n \, \partial w_1..\partial w_n} .$$

The relation

$$(2.5.11) \qquad P_j(u_1, u_2..u_j) = \int_{-\infty}^{\infty} \dots \int_{-\infty}^{\infty} P_n(u_1..u_j, u_{j+1}..u_n)\,du_{j+1} \dots du_n$$

is easily established from our former statements about the relation between distribution functions of different dimensionality. Eq.(2.5.11) implies the important fact that the information P_j is fully contained in the information given by P_n if $j < n$ and can easily be extracted from it.

In order to make the transition to the continuum we replace the set $u_1 \dots u_n$ by the field $u(x)$ and obtain the probability density functional $P\left[u(\underline{x})\right]$ or, more generally,

$P[u(\underline{x},t)]$. In case of a vector field $\underline{v}$ we have $P[\underline{v}(x)]$ and $P[v(\underline{x},t)]$ respectively. A discussion of these functionals will be given later.

2.6. Important average quantities

a) Expectations

The expectation of a random variable u is defined as the weighted mean of u. We write

$$E(u) \equiv \bar{u} \equiv \int_{-\infty}^{\infty} u\, P_1(u)\, du. \tag{2.6.1}$$

If u is a discrete variable then the integral in (2.6.1) is to be replaced by a sum. We shall not discuss this here.

For any function $g(u)$ of a random variable the expectation is defined as

$$E(g(u)) \equiv \overline{g(u)} \equiv \int_{-\infty}^{\infty} g(u)\; P_1(u)\, du. \tag{2.6.2}$$

In n dimensions we have

$$E(g(u_1, u_2 \ldots u_n)) = \overline{g(u_1, u_2 \ldots u_n)} \equiv$$

$$\equiv \int_{-\infty}^{\infty} \ldots \int_{-\infty}^{\infty} g(u_1, u_2 \ldots u_n) P_n(u_1, u_2 \ldots u_n) du_1 du_2 \ldots du_n. \tag{2.6.3}$$

b) Characteristic functions

The results of probability problems are often conveniently represented in Fourier space. The Fourier transforms of the density functions are called characteristic functions. In one dimension the characteristic function $M_1(v)$ is defined by

(2.6.4) $$M_1(\upsilon) \equiv E\left(e^{iu\upsilon}\right) = \int_{-\infty}^{\infty} e^{iu\upsilon} P_1(u)du.$$

For $\upsilon = 0$:

(2.6.5) $$M_1(0) = \int_{-\infty}^{\infty} P_1(u)du = 1.$$

We always have

(2.6.6) $$|M_1(\upsilon)| \leqslant M_1(0) = 1.$$

In n dimensions the characteristic function is

$$M_n(\upsilon_1, \upsilon_2 \ldots \upsilon_n) \equiv E\left(e^{i\sum_k \upsilon_k u_k}\right) =$$

(2.6.7) $$= \int_{-\infty}^{\infty} \ldots \int_{-\infty}^{\infty} e^{i\sum_k \upsilon_k u_k} P_n(u_1, u_2 \ldots u_n) du_1 du_2 \ldots du_n.$$

We always have

(2.6.8) $$M_n(0,0\ldots 0) = 1, \quad |M_n| \leqslant 1 ,$$

(2.6.9) $$M_n(\upsilon_1, \upsilon_2 \ldots \upsilon_{n-1}, 0) = M_{n-1}(\upsilon_1, \upsilon_2 \ldots \upsilon_{n-1}).$$

c) Moments of the form $\overline{u^n}$

The expressions

(2.6.10) $$\overline{u^n} \equiv E(u^n) = \int_{-\infty}^{\infty} u^n P_1(u)du$$

are called the <u>moments of the density function</u> $P_1(u)$. Particularly important are

$\bar{u}$ the 1. moment (= mean)

$\overline{u^2}$ the 2. moment .

The quantity

$$\sigma^2 = \overline{u^2} - \bar{u}^2 = \int_{-\infty}^{\infty} (u-\bar{u})^2 P_1(u)\,du \qquad (2.6.11)$$

is called the <u>variance</u> of $P_1(u)$. σ itself is the standard deviation. It is a measure of the width of the deviations from the mean. Another measure for this width is $\overline{|u'|}$ where

$$u' = u - \bar{u} \qquad (2.6.12)$$

is the deviation of u from the mean. Usually, $\overline{|u'|}$ and σ are close together. However, the analytical form of σ is more convenient for most calculations. The great importance of σ stems from this fact.

An important quantity, the Gauss probability density, is completely determined by the mean and the variance. In fact

$$P_{1G}(u) = \frac{1}{\sqrt{2\pi}\,\sigma}\, e^{-\left(\frac{u'}{\sigma}\right)^2} . \qquad (2.6.13)$$

The third-order moment $\overline{(u-\bar{u})^3}$ is a measure of the asymmetry of the density function around the mean.

If all moments up to the order ∞ are given then

the density $P_1(u)$ can be constructed. Of course, there exists a convergency condition for this to be true. We recognize this as follows: In $M_1(v)$ set

$$e^{iuv} = \sum_{n=0}^{\infty} \left(\frac{iuv}{n!}\right)^n. \tag{2.6.14}$$

Then

$$M_1(v) = \sum_{n=0}^{\infty} \frac{i^n v^n}{n!} \int_{-\infty}^{\infty} u^n P_1(u)\,du = \sum_{n=0}^{\infty} \frac{i^n v^n \overline{u^n}}{n!}. \tag{2.6.15}$$

If this power series in v converges absolutely for $v > 0$ then the Fourier transform of $M_1(v)$ does exist and is equal to $P_1(u)$. This consideration also shows that

$$\overline{u^n} = \frac{1}{i^n} \left. \frac{d^n M_1(v)}{dv^n} \right|_{v=0} \tag{2.6.16}$$

d) Correlation functions

So far we have considered only 1-dimensional moments. In two dimensions we need moments of the 2-dimensional density $P_2(u_1, u_2)$. These moments are defined as

$$\overline{u_1^j u_2^k} \equiv E(u_1^j u_2^k) = \int_{-\infty}^{\infty}\int_{-\infty}^{\infty} u_1^j u_2^k P_2(u_1, u_2)\,du_1 du_2. \tag{2.6.17}$$

Moments of this kind are also called 2-point correlation functions (sometimes 2-point coherence functions). We speak of "functions" because we allow $\underline{x}_1$ and $\underline{x}_2$ to be arbitrary variable points.

If $\underline{x}_1 = \underline{x}_2$ one also uses the term auto-correlation function and we use the term cross-correlation function if $\underline{x}_1 \neq \underline{x}_2$. If the complete set of these moments is given and obeys a suitable convergency condition then the 2-dimensional density P_2 can be constructed from the moments. Instead of (2.6.16) we now have

$$\overline{u_1^j u_2^k} = \frac{1}{i^{j+k}} \frac{\partial^j \partial^k}{\partial v_1^j \, \partial v_2^k} M_2(v_1, v_2) \Big|_{v_1 = v_2 = 0} . \qquad (2.6.18)$$

Of particular importance is the moment $\overline{u_1 u_2}$ often simply named the correlation function. It describes a certain relationship or correlation between the results of measurements u_1 (for instance at point x_1 and time t_1) and the results of measurements u_2 (for instance at point $\underline{x}_2$ and time t_2).

Also the moment $\overline{u_1^2 u_1^3}$, for example, describes such a correlation, although not quite the same. If there is no correlation between the outcomes in u_1 and u_2, then

$$\overline{u_1 u_2} = \bar{u}_1 \bar{u}_2 , \quad \overline{u_1^2 u_2^3} = \overline{u_1^2}\, \overline{u_2^3} \qquad (2.6.19)$$

and correspondingly for all other moments of this type. Hence the relation

$$\overline{u_1 u_2} = \bar{u}_1 \bar{u}_2 \qquad (2.6.20)$$

is necessary but not sufficient to ensure that the events measured by u_1 and u_2 respectively are independent. It is usual to call events which obey eq. (2.6.20) uncorrelated. Hence uncorrelated is not the same as independent. In practice, it means very often the same.

Expressed in terms of the diviation u' defined by eq. (2.6.12) we rewrite eq. (2.6.20) in the form

$$\overline{u_1' u_2'} = 0 \tag{2.6.21}$$

(condition for uncorrelated events).

Since the 2-point correlation function $\overline{u_1 u_2}$ is of such extreme importance it is often denoted by an extra letter. We write

$$R = R(\underline{x}_1, t_1; \underline{x}_2, t_2) \equiv \overline{u(x_1,t_1)u(x_2,t_2)} \,. \tag{2.6.22}$$

If $\underline{x}_1$ and $\underline{x}_2$ enter R in the special form $\underline{x}_1 - \underline{x}_2$ we say R is <u>homogeneous in space</u> or <u>stationary in space</u> or sometimes <u>statistically homogeneous</u>. If, on the other hand, t_1 and t_2 appear in the combination $t_1 - t_2$ only, then R is said to be <u>homogeneous in time</u> or <u>stationary in time</u>. The property of stationarity (in space or time), where it exists, implies a considerable simplification in all statistical calculations. The assumption of stationarity has been made in most of the successful non-trivial statistical treatments of continuum physical problems.

If $\underline{x}_1$ and $\underline{x}_2$ enter R in the special form $|\underline{x}_1 - \underline{x}_2|$ only then R is called to be <u>isotropic in space</u>.

Besides 2-point correlations also n-point correlations are needed in order to describe completely the interdependence of the results of measurements at the various space-time points. We define the general n-point correlation function by

$$\overline{u_1^a u_2^b \dots u_n^g} = \int_{-\infty}^{\infty} \dots \int_{-\infty}^{\infty} u_1^a u_2^b \dots u_n^g P_n(u_1, u_2 \dots u_n) du_1 du_2 \dots du_n . \tag{2.6.23}$$

It is plausible and can be proved that the complete statistical description of a system by means of statistical moments requires the full set of moments, i.e. the moments of all orders $n = 1, 2 \dots \infty$ (completeness theorem).

As mentioned above we allow the points $\underline{x}_1$, $\underline{x}_2$ etc. in the argument of the correlation functions to be arbitrary variable points. In particular, we also allow $\underline{x}_1 = \underline{x}_2$ in the 2-point correlation function $\overline{u_1 u_2} \equiv \overline{u(\underline{x}_1) u(\underline{x}_2)}$. In that case we obtain $\overline{u_1 u_2} = \overline{u_1^2}$. Similarly we have for instance

$$\overline{u^3(\underline{x}_1) u(\underline{x}_2) u^2(\underline{x}_3)} = \overline{u(\underline{x}_1) u(\underline{x}_2) u(\underline{x}_3) u(\underline{x}_4) u(\underline{x}_5) u(\underline{x}_6)} \tag{2.6.24}$$

in the special case $\underline{x}_1 = \underline{x}_4 = \underline{x}_5$, $\underline{x}_3 = \underline{x}_6$. This result implies that the above completeness theorem remains true if we remove the exponents $a, b \dots g$ in the definition (2.6.23) of the n-point correlation function.

We briefly mention a further kind of correlation functions. If v is a random variable distinct from u, i.e. measurements of u and v require different experiments, then $\overline{u_1 v_2}$ is called a <u>mixed correlation function</u> between u and v ; sometimes also the term cross-correlation function is used.

2.7. Ensemble theory

a) Stationary process

The notion of stationarity has been discussed already in connection with the 2-point correlation function. More generally, we call the n-dimensional probability density stationary in space and/or time if all its moments depend on $\underline{x}_1, \underline{x}_2 \ldots, t_1, t_2 \ldots$ via the relative quantities $\underline{x}_1 - \underline{x}_2$, $\underline{x}_1 - \underline{x}_3$, $\underline{x}_2 - \underline{x}_3$ etc. and/or $t_1 - t_2$, $t_2 - t_3$ etc. only. If the probability densities of all orders n are stationary then the whole problem is stationary in nature and we speak of stationary processes (in space or time).

Example: If a turbulent fluid flows through a tube of variable cross section then it has a varying mean velocity along the tube. If the difference in pressure at the ends of the tube is kept constant in time then one expects that the quantities measured macroscopically at a fixed arbitrary point do not vary in time: The flow is (macroscopically) stationary in time but not so in space. On the other hand, one can think of a turbulent flow which fills a certain part of space in a macroscopically homogeneous manner. If the turbulence decays uniformly in space, for instance due to the energy dissipated by the vortices, then the flow is stationary in space, but not in time. Problems of this kind are treated in detail by Batchelor in his book "The Theory of Homogeneous Turbulence".

b) <u>The notion of the ensemble</u>

In order to confirm experimentally the predictions of probability theory one needs a great number of experiments which must be of the same kind in their macroscopic performance but may be different in the microscopic details. For instance, the tossing of a die is such a macroscopic performance. The initial conditions, however, vary in a manner which is not well-known. These are the microscopic details. A set of such experiments is called an ensemble by Beran [B].

Instead, one can also think of an ensemble consisting of a large number of alike systems on which, macroscopically speaking, the same experiment is performed.

The ensemble theory is based on the following idea, which we illustrate by a particular example. In order to measure $\overline{u(\underline{x}_1,t_1)}$ as the mean value of the x-component of the velocity of a turbulent flow at a fixed point $\underline{x}_1$ in a tube one takes a large number of equal tubes, produces the same kind of flow in all of them (by opening the same sort of valves etc.), measures simultaneously the instantaneous velocity $u(\underline{x}_1,t_1)$ at corresponding points $\underline{x}_1$ and takes the average $\overline{u(x_1,t_1)}$ over the results in all tubes (*).

(*) In order to identify the average defined by the description given here with the mean previously defined with the help of $P_1(u_1)$ we have to give $P_1(u_1)$ a physical meaning. We prescribe that $P_1(u_1) = \lim p_n/n$ for $n\to\infty$ where p_n is the number of positive outcomes (here: in du_1 at u_1) in n trials.

If the flow is stationary in time then one expects the same results by taking the time average of u at $\underline{x}_1$ in a single tube rather than in many tubes provided that the time interval T chosen for the averaging is long enough. This average then can be written as

(2.7.1) $$\langle u(\underline{x}_1,t_1)\rangle_{t_1} \equiv \lim_{T\to\infty} \frac{1}{T}\int_T u(\underline{x}_1,t_1)dt_1 .$$

c) Ergodic hypothesis

The hypothesis that ensemble averaging and time averaging give the same result, i.e.

(2.7.2) $$\overline{u(\underline{x}_1,t_1)} = \langle u(\underline{x}_1,t_1)\rangle_{t_1}$$

is the ergodic hypothesis of our present (temporal) problem. Note, that the actual measurement, taken over a certain time interval which should be large compared to the characteristic time of the fluctuations, is closer to the time averaging procedure than to measurements on an ensemble. Hence we need an ergodic theorem if we want to use ensemble averages in our theory.

Similarly, we will be interested, as will often be the case in statistical continuum mechanics, in predicting certain quantities (e.g. stress, strain, velocity) in a certain region of space at a given time t_1, say. If $u(\underline{x}_1,t_1)$ represents this quantity then its expectation is the mean $\overline{u(\underline{x}_1,t_1)}$. The corresponding experiment again requires measurements of u over a great number of alike systems. If the situation is stationary in space then one expects the same result by taking the space average of u at t_1 in a single system, provided the region V of space chosen for the averaging is large enough. This average then can be written as

$$< u(\underline{x}_1, t_1) >_{\underline{x}_1} \equiv \lim_{V \to \infty} \frac{1}{V} \int_V u(\underline{x}_1, t_1) dV \qquad (2.7.3)$$

The hypothesis stating

$$\overline{u(\underline{x}_1, t_1)} = < u(\underline{x}_1, t_1) >_{\underline{x}_1} \qquad (2.7.4)$$

is the ergodic hypothesis of the present (spatial) problem.

There is a voluminous literature on the ergodic hypothesis justifying its use in all physically reasonably stated statistical problems (*). One should not forget, however the basic underlying assumption of stationarity. In actual problems where processes are never exactly stationary - since this requires infinite times or volumes - it is important to assure oneself that stationarity can be assumed approximately in space or time intervals which are large compared to the characteristic lengths or times of the fluctuations.

Example: In polycrystal problems the (spatial) ergodic theorem can be used only if the interval-mean quantities are nearly constant over distances which are large compared to the mean grain diameter. Notwithstanding, these distances can be very small compared to the external dimensions of the body, so that "macroscopic differentiation" has a meaning.

Little is known about the solution of statistical problems in which the ergodic hypothesis cannot be applied.

(*) See, for example, O. Onicescu, l.c.

2.8. Spectral representation

In the case of processes which are stationary in space and/or time, the 2-point correlation function R has one of the simplified forms

$$R(\underline{x}_1-\underline{x}_2, t_1-t_2),\ R(\underline{x}_1-\underline{x}_2, t_1, t_2),\ R(\underline{x}_1, x_2, t_1-t_2). \tag{2.8.1}$$

In this case one often uses with profit a Fourier representation of R . Let us consider a process which is stationary in time but not in space. If $\tau \equiv t_2-t_1$ then the Fourier transform in time of $R(\underline{x}_1, \underline{x}_2, \tau)$ is

$$S(\underline{x}_1, \underline{x}_2, \nu) = \int_{-\infty}^{\infty} R(\underline{x}_1, \underline{x}_2, \tau) e^{2\pi i\nu\tau} d\tau. \tag{2.8.2}$$

S is called cross spectral density if $\underline{x}_1 \neq \underline{x}_2$, and spectral density if $\underline{x}_1 = \underline{x}_2$. The term "density" refers to the scale of the frequencies ν .

To get a more intuitive understanding of the meaning of S we write, using the ergodic theorem in time (cf. eqs. 2.7.1, 2.7.2 of sect. 2.7) (*)

$$R(\underline{x}_1, \underline{x}_2, \tau) = \frac{1}{T}\int_T u(\underline{x}_1, t)u(\underline{x}_2, t+\tau)dt \equiv$$

$$\equiv \frac{1}{T}\int_T dt \int_{-\infty}^{\infty} \hat{u}(\underline{x}_1, \nu) e^{2\pi i\nu t} d\nu \int_{-\infty}^{\infty} \hat{u}(\underline{x}_2, \nu') e^{-2\pi i\nu'(t+\tau)} d\nu' =$$

$$= \frac{1}{T}\int_{-\infty}^{\infty} d\nu \int_{-\infty}^{\infty} d\nu'\ \hat{u}(\underline{x}_1, \nu)\hat{u}(\underline{x}_2, \nu') e^{-2\pi i\nu'\tau} \int_T dt\, e^{-2\pi i(\nu+\nu')t}. \tag{2.8.3}$$

(*) The following consideration also applies to complex u if the second u or $\hat{u}$ is always replaced by its complex conjugate.

In the limit $T \to \infty$ the last integral becomes $\delta(\nu + \nu')$ and we obtain after the performance of the ν'-integration

$$R(\underline{x}_1, \underline{x}_2, \tau) = \frac{1}{T} \int_{-\infty}^{\infty} \hat{u}(\underline{x}_1, \nu) \hat{u}(\underline{x}_2, -\nu) e^{+2\pi i \nu \tau} d\nu. \qquad (2.8.4)$$

Since R is the Fourier transform of S, we have the result that (cf. eq. 2.8.2)

$$S(\underline{x}_1, \underline{x}_2, \nu) = \frac{1}{T} \hat{u}(\underline{x}_1, -\nu) \hat{u}(\underline{x}_2, \nu). \qquad (2.8.5)$$

For $\underline{x}_1 = \underline{x}_2$ this is the celebrated formula of Wiener and Khinchin. There are problems of the existence of the limit of (2.8.5) for $T \to \infty$ for which the reader is referred to Beran ([B], p. 46). We shall assume that the limit does, in fact, exist in our physical problems of interest.

It follows from eq. (2.8.5) that $S(\underline{x}_1, \underline{x}_1, \nu)$ is proportional to the intensity with which the frequency ν occurs within the signal $u(\underline{x}_1, t)$.

The term "signal" refers to applications in electronics. A signal can be finite or infinite in time. Often it has a certain period. Then the corresponding amplitude appears to be especially large in the Fourier spectrum of the signal. This example demonstrates the utility of the spectral density.

In analogous manner one can deal with the processes which are stationary in space. If $\underline{w} = \underline{x}_2 - \underline{x}_1$, the spatial spectral density is

$$S(\underline{k},t_1,t_2) = \int_V R(\underline{w},t_1,t_2)e^{2\pi i\underline{k}\cdot\underline{w}}dV = \frac{1}{V}\hat{u}(-\underline{k},t_1)\hat{u}(\underline{k},t_2)$$
(2.8.6)

where $\hat{u}$ is now the space Fourier transform of u and V tends to infinity.

2.9. Functionals

Although we shall not use functionals in our later applications of the theory, they are of great importance in the formulation of continuum statistics. The point is that the complete set of moments which exhaustively describes the statistical situation of a continuous system, can be replaced by the probability density functional so giving the possibility for a very condensed formulation of the laws of continuum physics. It is to be expected that with the further development of the mathematical theory of functionals these will become increasingly important also in solving the statistical problems.

This section will be very brief. The interested reader is referred to the cited books of Volterra and of Evans.

We restrict ourselves to functionals in time which belong to a single function $u(t)$. The case of space functionals and of functionals which belong to several functions simultaneously can be treated in a similar way.

A time functional F is defined as a quantity which,

within our restriction, depends on the values of one function $u(t)$ of time. We assume that the time interval of interest ranges from a time a to a time b. We express the mentioned dependence by writing $F = F\left[u(t)\right]$.

It is convenient to think of functionals in terms of a limiting process for a function which depends on variables the number of which tends to infinity:

$$\lim_{n\to\infty} f(u_1, u_2 \ldots u_n) \Longrightarrow F\left[u(t)\right], \quad u_i \equiv u(t_i). \tag{2.9.1}$$

To illustrate this consider the simple case where

$$f_1(u_1 \ldots u_n) = \sum_{i=1}^{n} k_i u_i \tag{2.9.2}$$

is linear in u_i. The functional resulting for $n\to\infty$ then is of the form

$$F_1\left[u(t)\right] = \int_a^b k(t)\,u(t)\,dt. \tag{2.9.3}$$

If

$$u(t) = \lambda u_\lambda(t) + \mu u_\mu(t) \tag{2.9.4}$$

with two different functions u_λ and u_μ, then

$$F_1\left[u(t)\right] = \lambda F_1\left[u_\lambda(t)\right] + \mu F_1\left[u_\mu(t)\right]. \tag{2.9.5}$$

This means that the functional has maintained the property of linearity of the generating function f_1. Such functionals are called <u>linear</u>.

Similarly, a quadratic function

(2.9.6) $$f_2(u_1 \ldots u_n) = \sum_{i,j=1}^{n} k_{ij} u_i u_j$$

leads to a quadratic functional of the form

(2.9.7) $$F_2\left[u(t)\right] = \int_a^b \int_a^b k(t_1, t_2) u(t_1) u(t_2) dt_1 dt_2 .$$

Functionals of any degree can be formed in an analogous manner.

It is important to note that a rather general class of functionals can be written in the form

(2.9.8) $$G\left[u(t)\right] = \sum_{n=0}^{\infty} F_n\left[u(t)\right] .$$

This representation can be derived by expanding the generating fucntion f in a Taylor series.

Functionals can be differentiated and integrated. We refer the reader to the literature and only mention the interesting result: The n-th functional derivative of the characteristic functional $M\left[v(t)\right]$ at the place $v(t) = 0$ is i^n times the n-point correlation function. Here the characteristic functional is defined as the expectation of

(2.9.9) $$e^{i\int v(t)u(t)dt}$$

We recall that the characteristic function M_n was defined in section 2.6. as the expectation of

(2.9.10) $$e^{i\sum_{k=1} v_k u_k} ,$$

i.e. as

$$M_n(v_1, v_2 \ldots v_n) = \int_{-\infty}^{\infty} \ldots \int_{-\infty}^{\infty} e^{i \sum_{k=1}^{n} v_k u_k} P_n(u_1, u_2 \ldots u_n) du_1 du_2 \ldots du_n . \tag{2.9.11}$$

The characteristic functional follows from (2.9.11) by transition to the continuum:

$$M[v(t)] = \int e^{i \int v(t) u(t) dt} P[u(t)] du(t) . \tag{2.9.12}$$

where the integration is taken over the range of u .

CHAPTER III

The formation of statistical continuum mechanics

Except for a few physical examples, given just for illustration, everything said in chapter II is pure mathematics. Physical processes, stochastic or not, obey physical laws. Therefore any theory which belongs to the field of statistical physics must contain the concerned physical laws. We have mentioned this important fact already in the introduction.

Since we are interested in statistical continuum mechanics, the physical laws to be combined with the concepts of probability and statistics are those of continuum mechanics. The general procedure would be very similar if we develop a statistical continuum electrodynamics. Important insight into the essence of this general procedure can be gained if we consider the comparatively simple problem of the one-dimensional statistical harmonic oscillator. This problem is continuous in time whereas in continuum mechanics continuity in space is often more important. However, space or time, this does not make an essential difference in the structure of the theory.

3.10. the 1-dimensional statistical harmonic oscillator-functional formulation

a) Micro-time and macro-time information

The deterministic differential equation of the harmonic oscillator was written in section 1 to be of the form

$$m \frac{d^2u(t)}{dt^2} + k\,u(t) = f(t) \tag{3.10.1}$$

The meaning of the used symbols is m - mass, k - spring constant, f - exciting force, u - displacement. u is the random variable about which we would like to make predictions. The stochastic character of u can be due to random behaviour in time of m and/or k and/or f. Let us consider the case of a random driving force $f(t)$ with constant m and k. There are many physical examples of such a force. To give just one example: in some approximation one may treat an aircraft wing as an oscillator which experiences random forces by a turbulent air.

Assuming we know $f(t)$ in all detail, we just need to solve eq. (3.10.1) with the given initial conditions - for instance u and du/dt given at time 0 - in order to obtain u as function of t. This $u(t)$ together with $f(t)$ would contain the complete deterministic information on the oscillator in this case. We shall call this information the micro-time information because it is an information on the time scale of the fluctuations i.e. on a very small time scale.

If the problem is really stochastic in nature,

then the function $f(t)$ will be so complicated, that one would need computers of largest capacities in order to obtain a reasonably good solution over any macro-time interval which, by definition, is large compared to the characteristic fluctuation time. In practice, one is usually neither interested in the mentioned details of the driving force nor in those of the displacement. It is rather the average behaviour of these quantities, i. e. the behaviour on the macro-time scale which is of interest.

In order to see what is needed for such an average description consider the random function $f(t)$. All possible ensemble averages involving $f(t)$ contain average, i.e. macro-time, information about $f(t)$. Such averages are the mean $\overline{f(t_1)}$, the 2-point correlation function $\overline{f(t_1)f(t_2)}$, the n-point correlation function $\overline{f(t_1)f(t_2)\ldots f(t_n)}$ etc., where $t_1, t_2 \ldots t_n$ are continuous variables. The complete statistical information about f is, at the same time, the complete macro-time information about f. This is given by the complete set of correlation functions $(n = 1,2 \ldots \infty)$. As pointed in section 2.5. this set is equivalent to the complete set of f-moments of the probability density functions of order $n = 1,2 \ldots \infty$. From the set of f-moments one can construct the density functions themselves, and their complete set again contains the same amount of information as the probability density functional of f. In other words the probability density functional $P[f(t)]$ provides the complete statistical, i.e. macro-time information on the random function f . An analogous statement is true for any other random function.

b) Formulation of the problem

Although a functional is, in general, a more complicated entity than a function, this is not true in the present case because $f(t)$ is a function on the micro-scale whereas the functional $P[f(t)]$ refers to the macro-scale. This corresponds to the obvious fact that the functional P contains less information than the function f .

We can now provisionally formulate the mathematical problem of the statistical harmonic oscillator with random driving force: Given the probability density functional $P[f(t)]$ of the driving force $f(t)$, find the probability density functional $P[u(t)]$ of the displacement $u(t)$.

It is important to note that $P[f(t)]$ and $P[u(t)]$ do not contain the whole macro-time information about the oscillator. In fact, the information missing is about the mixed probability functional $P[u(t), f(t)]$ or the corresponding mixed correlation functions of u and f , for instance $\overline{u(t_1)f(t_2)}$, $\overline{u(t_1)u(t_2)f(t_3)u(t_4)}$ etc. which are formed with the mixed density functions $P_2(u_1, f_2)$, $P_4(u_1, u_2, f_3, u_4)$ etc. respectively. The additional information is about the correlation between $u(t)$ and $f(t)$. It answers, for instance, the question whether u is always large at a time when f is large or whether there is a definite time shift between large u and large t etc.. Of course such information can be of great interest.

Since $P[u(t), f(t)]$ contains both $P[u(t)]$ and $P[f(t)]$ we can, finally, formulate the mathematical problem as follows: Given the force functional $P[f(t)]$, find the mixed force-displacement functional $P[u(t), f(t)]$.

c) The governing functional equation

We expect that the desired functional obeys a functional equation which can be derived from the deterministic oscillator equation (3.10.1). There exists a definite procedure to do this which is described by Beran [B]. The resulting functional equation is obtained not for the probability density functional P, but rather for the corresponding characteristic functional $M = M[v(t), g(t)]$ where g is conjugate to f in the same sense as v is to u. Just to provide an idea of the final mathematical problem we give the result of this calculation. The functional M has to be found from the functional equation

$$m \frac{d^2}{d\xi^2} \frac{\delta M[v(t), g(t)]}{\delta v(\xi)} + k \frac{\delta M[v(t), g(t)]}{\delta v(\xi)} = \frac{\delta M[v(t), g(t)]}{\delta g(\xi)} \tag{3.10.2}$$

under the "boundary condition"

$$M[v(t), g(t)] = M[g(t)] \quad \text{for } v(t) = 0. \tag{3.10.3}$$

In addition, initial values can be prescribed for u and du/dt. In eq. (3.10.2) $\delta/\delta v(\xi)$ and $\delta/\delta g(\xi)$ symbolize the functional derivatives with respect to $v(\xi)$ and $q(\xi)$ respectively. This equation shows some resemblance to the original oscillator equation (3.10.1).

Functional analysis is not yet developed sufficiently to make it promising to attack eq. (3.10.2) directly. The lucidity and succinctness of the functional language makes it particularly suitable for theoretical considerations of the statistical continuum problems. It has, however, not yet reached any importance in the field of applications. Here the formulation in terms of moments, i.e. of correlation functions, is of much greater relevance.

3.11. The 1-dimensional statistical harmonic oscillator-moment formulation

By means of the result mentioned in section 2.9. one can derive the equations governing the statistical moments directly from the functional equations [B]. We shall not do this but instead obtain the moment equations by some operations on the original deterministic equation.

The moments for which we seek governing equations are the correlation functions of $u(t)$ and the mixed correlation functions of $u(t)$ and $f(t)$. Thus these moments are of the general form

$$u(t_1)\ldots u(t_j)\,f(t_{j+1})\ldots f(t_n) \qquad (3.11.1)$$

The correlation functions of $f(t)$ are considered as the given quantities.

In order to derive the desired equations we need a rule for interchanging the process of differentiation and ensemble averaging. Let the members be labelled by a number ν which goes from 1 to a very big number, practically to infinity. The ensemble average $\overline{u(t)}$ of $u(t)$ is then defined as

$$\overline{u(t)} = \frac{1}{N} \sum_{\nu=1}^{N} u_\nu(t) \tag{3.11.2}$$

where $u_\nu(t)$ is the value of u measured in the ν-th ensemble at time t. Now

$$\frac{d}{dt}\overline{u(t)} = \frac{1}{N} \sum_{\nu=1}^{N} \frac{d}{dt} u_\nu(t) = \overline{\frac{du(t)}{dt}}. \tag{3.11.3}$$

So we have obtained the result that differentiation and ensemble averaging can be interchanged, a result of fundamental importance.

In order to derive the moment equations we first rewrite the oscillator equation in terms of t_1 instead of t :

$$m \frac{d^2u(t_1)}{dt_1^2} + k\, u(t_1) = f(t_1). \tag{3.11.4}$$

We then multiply this by all possible factors of the form

$$u(t_2) \ldots u(t_m) f(t_{m+1}) \ldots f(t_p) \tag{3.11.5}$$

form the ensemble mean and interchange differentiation with averaging. The lowest-dimensional terms of the form (3.11.5) are listed below where we have substituted $u(t_i) \equiv u_i$, $f(t_i) \equiv f_i$:

$$
\begin{gathered}
1 \\
f_2 \;,\; u_2 \\
f_2 f_3 \;,\quad f_2 u_3 \;,\quad u_2 u_3 \\
f_2 f_3 f_4 \;,\; f_2 f_3 u_4 \;,\; f_2 u_3 u_4 \;,\; u_2 u_3 u_4 \\
\ldots \ldots \ldots
\end{gathered}
\tag{3.11.6}
$$

In the respective sequence we obtain the governing equations for the correlation functions of order 1, 2, 3, 4 ..., namely

$$
m \frac{d^2 \overline{u}_1}{dt_1^2} + k \overline{u}_1 = \overline{f}_1 \qquad \text{1st order}
$$

$$
\left.
\begin{aligned}
m \frac{\partial^2 \overline{u_1 f_2}}{\partial t_1^2} + k\, \overline{u_1 f_2} &= \overline{f_1 f_2} \\
m \frac{\partial^2 \overline{u_1 u_2}}{\partial t_1^2} + k\, \overline{u_1 u_2} &= \overline{f_1 u_2}
\end{aligned}
\right\} \quad \text{2nd order}
$$

$$
\left.
\begin{aligned}
m \frac{\partial^2 \overline{u_1 f_2 f_3}}{\partial t_1^2} + k\, \overline{u_1 f_2 f_3} &= \overline{f_1 f_2 f_3} \\
m \frac{\partial^2 \overline{u_1 f_2 u_3}}{\partial t_1^2} + k\, \overline{u_1 f_2 u_3} &= \overline{f_1 f_2 u_3} \\
m \frac{\partial^2 \overline{u_1 u_2 u_3}}{\partial t_1^2} + k\, \overline{u_1 u_2 u_3} &= \overline{f_1 u_2 u_3}
\end{aligned}
\right\} \quad \text{3rd order}
\tag{3.11.7}
$$

. . . .

Observe that the correlations of different order are not coupled by these equations. This is a special result because the oscillator equation we have used is a linear equation with constant coefficients. It can be seen immediately that a coupling of equations for moments of different orders occurs if we introduce a non-linear term into the oscillator equation. The various equations within a particular order are coupled but in a simple way. For instance, having found $\overline{u_1 f_2 f_3}$ from the first 3rd-order equation we can calculate $\overline{u_2 f_2 u_3}$ from the second and then $\overline{u_1 u_2 u_3}$ from the third equation.

Although the correlation functions calculated from eqs. (3.11.7) form the formal solution they are not always the physical quantities of immediate interest. In some cases the characteristic functions, for instance, are closer to the physical fact. One can easily obtain the equations governing the latter functions as the Fourier transforms of eqs. (3.11.7). We shall show this later in the example of turbulent flow.

3.12. The equations of motion of statistical continuum mechanics

a) The general deterministic equations of motion

With the knowledge of the preceding sections we can now proceed rather quickly. The deterministic equations of motion of continuum mechanics are

$$\rho \frac{d\underline{v}}{dt} + \operatorname{div} \underline{\sigma} + \rho \underline{F} = 0 \qquad (3.12.1)$$

where $\varrho(\underline{x},t)$ denotes the mass density, $\underline{v}(\underline{x},t)$ the velocity field, $\underline{\sigma}(\underline{x},t)$ the symmetric stress tensor, and $\underline{F}(\underline{x},t)$ the external forces per unit mass. The field quantities of interest are $\underline{v}$ and $\underline{\sigma}$.

Eq. (3.12.1) has some resemblance with the oscillator equations as it contains a material property - the density - giving rise to inertia; it further includes driving forces $\underline{F}$; finally, $\mathrm{div}\,\underline{\sigma}$ depends on a material property - elasticity or viscosity - which is more or less analogous to the spring constant k in the oscillator equation. We then expect a stochastic behaviour of our continuum if ϱ and/or $\underline{F}$ and/or the quantities characterizing the elasticity or viscosity of the continuum are random quantities in space and/or time.

It can easily be shown (do it yourself!) that the total derivation d/dt cannot be interchanged with the averaging because of the appearance of $\underline{v}$ in $d/dt = \partial/\partial t + \underline{v}\cdot\nabla$ whereas this is possible with the partial derivations $\partial/\partial t$ and $\partial_i \equiv \partial/\partial x_i$.

Hence we shall not apply the procedure, used in the example of the harmonic oscillator, directly to eq. (3.12.1) but to the more developed form

$$-\varrho(\underline{x}_1,t_1)\frac{\partial\underline{v}(\underline{x}_1,t_1)}{\partial t_1} - \varrho(\underline{x}_1,t_1)\,\underline{v}(\underline{x}_1,t_1)\cdot\underline{\nabla}_1\underline{v}(\underline{x}_1,t_1) + $$
$$+\underline{\nabla}_1\cdot\underline{\sigma}(\underline{x}_1,t_1) + \varrho(\underline{x}_1,t_1)\,\underline{F}(\underline{x}_1,t_1) = 0 \qquad (3.12.2)$$

Here we have changed the independent variables into $\underline{x}_1, t_1$. Eq.

(3.12.2) governs the behaviour of the continuum on the (space and time) micro-scale.

b) Dynamics of incompressible Newtonian fluids

If we limit ourselves to incompressible fluids we have in addition to eq. (3.12.2)

$$\underline{\nabla}_1 \cdot \underline{v}(x_1, t_1) = 0 . \tag{3.12.3}$$

We further assume the isotropic Newton's law

$$\sigma_{ij} - p\,\delta_{ij} = \varrho\nu\,(\partial_i v_j + \partial_j v_i) \tag{3.12.4}$$

(in cartesian coordinates). Here p is the hydrodynamic pressure and $\varrho\nu$ the constant of viscosity. Using (3.12.4) in (3.12.2) we obtain Navier-Stokes' equations in the form

$$\frac{\partial \underline{v}(\underline{x}_1, \underline{t}_1)}{\partial t_1} + \underline{v}(\underline{x}_1, \underline{t}_1) \cdot \underline{\nabla}_1 \underline{v}(\underline{x}_1, t_1) + \frac{1}{\varrho}\,\underline{\nabla}_1 p(\underline{x}_1, t_1) -$$

$$- \nu\,\underline{\nabla}^2 \underline{v}(\underline{x}_1, t_1) - \underline{F}(\underline{x}_1, t_1) = 0 . \tag{3.12.5}$$

These equations, together with (3.12.3) have, among others, solutions which fluctuate on a small scale. This is true even if ϱ, ν and $\underline{F}$ are deterministic quantities. Of course, we now speak of the important phenomenon of turbulence which can be described only by non-linear equations. To see how the general method of section 3.11 works in our present situation, let us establish the equations which govern the 2-point correlation functions in the special case $F = 0$. We first observe that due to eq. (3.12.3)

$$\underline{v} \cdot \underline{\nabla}_1 \underline{v} = \underline{\nabla}_1 \cdot (\underline{v}\,\underline{v}) .$$

Having substituted this in eq. (3.12.5) we multiply by $\underline{v}(\underline{x}_2, \underline{t}_2)$ and take the average. Interchanging then averaging with differentiation and using now subscript notation in Cartesian coordinates we arrive at

$$\frac{\partial}{\partial t_1} \overline{v_i(\underline{x}_1,t_1)\, v_k(\underline{x}_2,t_2)} + \frac{\partial}{\partial x_{1j}} \overline{v_j(\underline{x}_1,t_1)\, v_i(\underline{x}_1,t_1)\, v_k(\underline{x}_2,t_2)} +$$

$$+ \frac{1}{\varrho} \frac{\partial}{\partial x_{1i}} \overline{p(\underline{x}_1,t_1)\, v_k(\underline{x}_2,t_2)} - \nu \nabla_1^2 \overline{v_i(\underline{x}_1,t_1)\, v_k(\underline{x}_2,t_2)} = 0 .$$

(3.12.7)

In a similar way we obtain the equation for the correlation $v_i(\underline{x}_1, t_1) p(\underline{x}_2, t_2)$. The result is

$$\frac{\partial}{\partial t} \overline{v_i(\underline{x}_1,t_1)\, p(\underline{x}_2,t_2)} + \frac{\partial}{\partial x_{1j}} \overline{v_j(\underline{x}_1,t_1)\, v_i(\underline{x}_1,t_1)\, p(\underline{x}_2,t_2)} +$$

$$+ \frac{1}{\varrho} \frac{\partial}{\partial x_{1i}} \overline{p(\underline{x}_1,t_1)\, p(\underline{x}_2,t_2)} - \nu \nabla^2 \overline{v_i(\underline{x}_1,t_1)\, p(\underline{x}_2,t_2)} = 0 .$$

(3.12.8)

In addition to eqs. (3.12.7), (3.12.8) the divergence conditions

$$\frac{\partial}{\partial x_{1i}} \overline{v_i(\underline{x}_1,t_1)\, v_k(\underline{x}_2,t_2)} = 0 , \quad \frac{\partial}{\partial x_{1i}} \overline{v_i(\underline{x}_1,t_1)\, p(\underline{x}_2,t)} = 0$$

(3.12.9)

must be satisfied. The important point is that 3rd-order functions appear in the equations for the 2nd-order correlation functions. In order to calculate these we first should know the 3rd-order functions. If we set up their equations then 4th-order equations appear etc.. In this way the problem becomes very complex, in fact, unsolvable in a strict sense. Notwithstanding, we shall see in section 5.18, how one can obtain physically sensible solutions in certain situations.

c) Linear statistical elastodynamics

In the linearized theory the total derivatives in the equation of motion can be replaced by the partial derivatives in time. We assume Hooke's law in the general anisotropic form

$$\sigma_{ij} = c_{ijkl}\,\partial_k u_l \tag{3.12.10}$$

where u_l is the displacement field and obtain the deterministic equation

$$-\varrho(\underline{x},t)\,\frac{\partial^2 u_j(\underline{x},t)}{\partial t^2} + \frac{\partial}{\partial x_i}\left[c_{ijkl}(\underline{x},t)\,\frac{\partial u_l(\underline{x},t)}{\partial x_k}\right] + \varrho F_j(\underline{x},t) = 0 \tag{3.12.11}$$

The stochastic properties enter via ϱ, c_{ijkl} and/or F_j . Since it is now clear how to obtain the equations for statistical moments we need not write them down. We shall later treat in detail the case of random distribution of the elastic moduli c_{ijkl} in the static case without external forces. Our (deterministic) starting equation will then be simply

$$\frac{\partial}{\partial x_i}\left[c_{ijkl}(\underline{x},t)\,\frac{\partial u_l(\underline{x},t)}{\partial x_k}\right] = 0\,. \tag{3.12.12}$$

3.13. Use of Green's functions

In linear theories one can often write down the solution by use of a Green's function. Thus the general solution of the oscillator equation, for example, can be written in the form

$$u(t_1) = \int_{-\infty}^{t_1} G(t_1, t_2) f(t_2) dt_2 \qquad (3.13.1)$$

with G as the Green's function which can be adapted to the special initial conditions of the problem. We assume that the Green's function is a random function by taking in the oscillator equation m and/or k as random in time.

With the representation (3.13.1) of u we can form correlation functions of any kind, for instance

$$\overline{u(t_1)u(t_2)} = \overline{\int_{-\infty}^{t_1}\int_{-\infty}^{t_2} G(t_1,t_3)\, G(t_2,t_4)\, f(t_3)\, f(t_4)\, dt_3\, dt_4} \,. \qquad (3.13.2)$$

If we assume a stationary situation then we can extend the integrals to infinity. We also assume that the ergodic theorem applies. Then we have with $\tau \equiv t_2 - t_1$

$$\overline{u(t_1)u(t_2)} = \langle u(t_1)u(t_2)\rangle_t = \frac{1}{T}\int_T u(t_1)u(t_1+\tau)dt_1 \,. \qquad (3.13.3)$$

This integration in which τ is kept constant should be interchangeable with those of eq. (3.13.2) so that we obtain

$$\overline{u(t_1)\,u(t_2)} = \int_{-\infty}^{\infty}\int_{-\infty}^{\infty} < G\,(t_1,t_3)\,G\,(t_2,t_4)>_t f(t_3)f(t_4)dt_3dt_4 .$$

(3.13.4)

We denote the correlation function on the left by $\upsilon(\tau)$. Since we are considering a stationary problem the correlation function of G which we call Γ/T depends on t_1, t_2 only in the combination τ. We expect that due to the symmetry of the Green's function Γ contains t_3, t_4 only in the combination $\tau' = t_4 - t_3$. Hence $\Gamma = \Gamma(\tau,\tau')$ and this is a macro-time function. The integrations in eq. (3.13.4) can be rewritten as integrations over t_3 and τ' if we write $f(t_4) = f(t_3 + \tau')$. Since τ' is to be kept constant during the t_3-integration this can be performed and results in $T < f(t_3)f(t_4)>_t$, which is a function of τ' alone. If we denote the f-correlation by $\varphi(\tau')$ then we have finally obtained

$$\upsilon(\tau) = \int_T \Gamma(\tau,\tau')\,\varphi(\tau')\,d\tau' . \tag{3.13.5}$$

This result is very much simpler than the solution (3.13.1) and correspondingly contains less information. In fact, (3.13.1) is the micro-scale solution whereas (3.13.5) is a part of the macro-scale solution. Of course, the question of practical importance is whether we can obtain $\Gamma(\tau,\tau')$ easily. In relevant cases this is simpler than to calculate $G(t_1,t_2)$ because we need less information for $\Gamma(\tau,\tau')$.

We have obtained the 2-time-correlation function

without formulating an equation for it. This shows the power of the Green's function method which is quite important also in many other areas of physics. In chapter V we shall use Green's functions in order to solve the problem of the effective elastic moduli of bodies with a heterogeneous constitution.

Complications can arise if the situation is not stationary in time. If the variation of the average quantities in time is slow enough to ensure the validity of the ergodic theorem (see section 2.7.) then one can proceed along similar lines also in the non-stationary case.

Everything said so far on the Green's function method for the time problem applies in a corresponding form to the spatial problem.

CHAPTER IV

The statistical problem of turbulence

4.14. Formulation of the basic equations in ordinary space

a) Statement of the physical problem

This is the classical example of an application of statistical and probability theory in continuum mechanics. The stochastic nature of the turbulent flow follows from the fact that the particle velocities fluctuate over distances which are small compared to other dimensions of the flow. The question whether turbulent flow is a solution of the Navier-Stokes equations has been discussed and is not trivial. Today it seems to be generally accepted that the Navier-Stokes equations should, in fact, allow turbulent flow.

Of course, the Navier-Stokes equations also describe laminar flow. Whether a laminar or turbulent flow is stable in a certain situation is not a problem of statistics but a stability problem which will not be considered here. To a certain extent this question has been answered by O. Reynolds in his early work on the problem of turbulence.

To obtain solutions of the deterministic Navier-Stokes equations (eqs. 3.12.5) is impossible because these would be solutions on the micro-scale, hence extremely complicated.

The methods described in chapter III allow us to formulate moment equations on the macro-scale. The particular difficulty with these equations is that the 2nd-order correlations can be obtained from them only if the 3rd-order correlations are known, and the determination of these requires the knowledge of 4th-order correlations etc.. For this reason it is also impossible, to obtain solutions of the macro-scale equations. It is, however, possible to find solutions if one makes certain assumptions on the higher order correlations, and it is this fact which has lead to some success in the statistical theory of turbulence.

Before starting with the theory proper it might be desirable to list some of the main contributors to the statistical approach of turbulence. This list is by no means complete. We refer to G.I. Taylor (1921), A.A. Friedmann and A. Keller (1924), Th. von Karman (1937), A. Kolmogorov (1941), C.F. von Weizsäcker (1948); W. Heisenberg (1948), E. Hopf (1952). Further references are found in Beran [B] or given at the end of this text.

We assume that the phenomenon of turbulence as such is known to the reader. We shall start from the deterministic Navier-Stokes equations. For simplicity we assume that no external forces are present and that the density ϱ is constant in space and time. In other words, we investigate the flow of an incompressible fluid. We shall in addition require that the mean flow $\overline{\underline{v}(\underline{x},t)}=0$. As an example we mention a fluid in a basin which is excited somehow, for instance by a certain machinery, to a turbulent situation. Having once produced the turbulence, one can think of either sustaining it or letting it decay by its own dissipation of energy. In order to avoid boundary complications we assume the space filled by the fluid to be infinite in extension.

Beside the above mentioned simplifications we shall have to introduce later on further idealizations in order to arrive at definite results. These are the assumptions of stationarity in space and time and of macroscopic isotropy.

b) Derivation of the basic equation

Throughout this text we denote deviations from the mean by a prime. Since the mean of the velocity vector field is zero by assumption, we have $\underline{v} = \underline{v}'$. Hence we shall omit the prime on $\underline{v}$. With $\underline{F} = 0$ and constant ϱ we now obtain the Navier-Stokes equations (eqs. 3.12.5) in the form

$$\frac{\partial}{\partial t} v_i(\underline{x}_1, t) + \frac{\partial}{\partial x_{1j}} \left[v_i(\underline{x}_1, t) v_j(\underline{x}_1, t) \right] =$$

$$= \nu \underline{\nabla}_1^2 v_i(\underline{x}_1, t) - \frac{1}{\varrho} \frac{\partial p(\underline{x}_1, t)}{\partial x_{1i}} \tag{4.14.1}$$

which applies together with the continuity equation

$$\frac{\partial}{\partial x_{1j}} v_j(\underline{x}_1, t) = 0 \,. \tag{4.14.2}$$

We have not substituted t_1 for t because we do not intend to form correlation functions in time but rather in space. In applying this restriction which serves to simplify the situation we follow many of the authors who have been concerned with the present problems.

Proceeding as in chapter III we multiply by $v_k(\underline{x}_2, t)$ take the ensemble average and interchange this with the differentiation. We thus obtain

$$\frac{1}{2}\left\{\frac{\partial}{\partial t}\overline{v_i(\underline{x}_1,t)v_k(\underline{x}_2,t)} + \frac{\partial}{\partial x_{1j}}\overline{v_i(\underline{x}_1,t)v_j(\underline{x}_1,t)v_k(\underline{x}_2,t)} - \right.$$

$$\left. -\nu\nabla_1^2\overline{v_i(\underline{x}_1,t)v_k(\underline{x}_2,t)} + \frac{1}{\varrho}\frac{\partial}{\partial x_{1i}}\overline{p(\underline{x}_1,t)v_k(\underline{x}_2,t)}\right\} + \left\{i1 \leftrightarrow k2\right\} = 0 \tag{4.14.3}$$

where $\left\{i1 \leftrightarrow k2\right\}$ denotes the expression obtained by the simultaneous substitution $i \leftrightarrow k$, $1 \leftrightarrow 2$.

This equation is to be supplemented by divergency conditions which follow from eq. (4.14.2).

The eqs. (4.14.3) are rigorous equations under the stated restrictions $(\overline{\underline{v}} = 0, \underline{F} = 0, \varrho = \text{const.})$. Together with the divergency conditions they would be sufficient in number to calculate the 2nd-order correlations $\overline{v_i v_k}$ and $\overline{p v_k}$, if the 3rd-order correlations in (4.14.3) were known functions. The actual lack of information will be compensated later by an assumption inferred from physical intuition.

c) <u>Stationarity in space (homogeneous turbulence)</u>

"The Theory of Homogeneous Turbulence" is the title of Batchelor's book which is highly recommended to the reader with special interest in the turbulence problem. Stationarity in space implies that the correlation functions depend on difference vectors only. Using the notation of Beran [B] which is the same as that of Batchelor we define

$$R_{ij}(\underline{r},t) \equiv \overline{v_i(\underline{x}_1,t)v_j(\underline{x}_2,t)} \tag{4.14.4}$$

$$S_{ijk}(0,\underline{r},t) \equiv S_{ijk}(\underline{r},t) \equiv \overline{v_i(\underline{x}_1,t)v_j(\underline{x}_1,t)v_k(\underline{x}_2,t)} \tag{4.14.5}$$

$$P_j(\underline{r},t) \equiv \frac{1}{\varrho}\,\overline{p(\underline{x}_1,t)v_j(\underline{x}_2,t)} \tag{4.14.6}$$

where $\underline{r} \equiv (r_i) \equiv \underline{x}_2 - \underline{x}_1$.

Since $\partial/\partial x_{1i} = -\partial/\partial r_i$, we derive from eq. (4.14.3)

$$\left\{\frac{1}{2}\frac{\partial}{\partial t}R_{ik}(\underline{r},t) - \frac{\partial}{\partial r_j}S_{ijk}(\underline{r},t) - \nu\,\underline{\nabla}_r^2 R_{ik}(\underline{r},t) + \right.$$

$$\left. + \frac{\partial}{\partial r_i}P_k(\underline{r},t) + \{i1 \leftrightarrow k2\}\right\} = 0. \tag{4.14.7}$$

From eq. (4.14.2) one easily finds the following divergency conditions on R , S and P :

$$\frac{\partial R_{ik}}{\partial r_i} = 0\,, \quad \frac{\partial S_{ijk}}{\partial r_k} = 0\,, \quad \frac{\partial P_i}{\partial r_i} = 0\,. \tag{4.14.8}$$

Eqs. (4.14.7 - 4.14.8) are the basic equations of our problem formulated in ordinary space. They are exact under the stated restrictions. However, they contain only a limited macroscopic information since from them only 2nd-order correlations can be calculated.

We could simplify the eqs. (4.14.7 - 4.14.8) further by assuming macroscopic isotropy as done by Taylor and by von Kármán and Howarth. It turns out, however, that it is more convenient to introduce this restriction after the transition to the Fourier space. This transition is now very simple because with the assumption of stationarity in space the equations of the ordinary space depend on the difference vector $\underline{r}$, but not on $\underline{\mathfrak{x}}_1$ and $\underline{\mathfrak{x}}_2$ separately.

4.15. Transition to Fourier space. Final results

a) Fourier transformation

We introduce the Fourier transforms of the correlations R, S, P by

$$\Phi_{ij}(\underline{k},t) = \frac{1}{(2\pi)^3}\int R_{ij}(\underline{r},t)e^{-i\underline{k}\cdot\underline{r}}dV \qquad (4.15.1)$$

$$\Upsilon_{ijk}(\underline{k},t) = \frac{1}{(2\pi)^3}\int S_{ijk}(\underline{r},t)e^{-i\underline{k}\cdot\underline{r}}dV \qquad (4.15.2)$$

$$\Theta_i(\underline{k},t) = \frac{1}{(2\pi)^3}\int P_i(\underline{r},t)e^{-i\underline{k}\cdot\underline{r}}dV \qquad (4.15.3)$$

Eq. (4.14.7) then becomes

$$\frac{1}{2}\left\{\frac{\partial\Phi_{ik}(\underline{k},t)}{\partial t} - ik_j\Upsilon_{ijk}(\underline{k},t) + \right.$$

$$\left. + \nu k^2\Phi_{ik}(\underline{k},t) + ik_i\Theta_k(k,t)\right\} + \{\ \} = 0, \qquad (4.15.4)$$

where the meaning of $\{\ \}$ from (4.14.7), and the divergency conditions become

$$k_i\Phi_{ik} = 0,\quad k_k\Upsilon_{ijk} = 0,\quad k_i\Theta_i = 0 \qquad (4.15.5)$$

Physically we now consider waves of wavelength λ, wave vector $\underline{k}$ and wave number $k \equiv |\underline{k}| = 2\pi/\lambda$. We expect that the intensity of long waves is large when the fluid contains many large elements of turbulence (eddies), whereas small elements of turbulence contribute to the short waves.

b) Macroscopic isotropy

This is a further important simplification. As shown by Batchelor in his book the tensors Φ , Υ and Θ must now be of the following form:

$$\Phi_{ik}(\underline{k},t) = a(k,t)\delta_{ik} + b(k,t)k_i k_k \tag{4.15.6}$$

$$\Upsilon_{ijk}(\underline{k},t) = a_1(k,t)k_i k_j k_k + a_2(k,t)(k_i\delta_{jk} + k_j\delta_{ik}) + a_3(k,t)k_k\delta_{ij} \tag{4.15.7}$$

$$\Theta_j(\underline{k},t) = 0 . \tag{4.15.8}$$

The functions a, b, a_1, a_2, a_3 depend on $\underline{k}$ only via $k \equiv |\underline{k}|$.

Eqs. (4.15.6 - 4.15.8) stipulate that

$$a(k,t) + k^2 b(k,t) = 0 \tag{4.15.9}$$

$$a_3(k,t) = 0, \quad k^2 a_1(k,t) + 2a_2(k,t) = 0 \tag{4.15.10}$$

Hence the Φ_{ik} and Υ_{ijk} can each be expressed by one single scalar function in such a way that the divergency conditions are satisfied identically. Using Beran's notation we write

$$\Phi_{ik}(\underline{k},t) = \frac{E(k,t)}{4\pi k^4}(k^2\delta_{ij} - k_i k_k) \tag{4.15.11}$$

$$\Upsilon_{ijk}(k,t) = i\,\Upsilon(k,t)\Big[k_i k_j k_k - \\ - \frac{1}{2}k^2\left(k_i\delta_{jk} + k_j\delta_{ik}\right)\Big]. \qquad (4.15.12)$$

Substitution of these relations and $\Theta_k = 0$ in eq. (4.15.4) yields after simple calcuation (*)

$$\frac{\partial E(k,t)}{\partial t} + 2\nu k^2 E(k,t) = 4\pi k^6 \Upsilon(k,t). \qquad (4.15.13)$$

This is the basic equation of the homogeneous and isotropic turbulence in Fourier space, as far as one is concerned with 2nd-order correlations. $E(k,t)$ is essentially the Fourier transform of the correlation function $\overline{v_1 v_2}$; hence it is essentially the spectral density of $\underline{v}(\underline{x},t)$. Eq. (4.15.13) is rigorous.

In order to give a physical interpretation of E let us take the trace of eq. (4.15.11):

$$E(k,t) = 2\pi k^2 \Phi_{ii}(\underline{k},t). \qquad (4.15.14)$$

On the other hand

$$e_{kin}(t) = \frac{\varrho}{2}R_{ii}(0,t) = \frac{\varrho}{2}\overline{v_i(\underline{x}_1,t)v_i(\underline{x}_1,t)} =$$

$$= \frac{\varrho}{2}\int \Phi_{ii}(\underline{k},t)e^{i\underline{k}\cdot\underline{r}}d^3\underline{k}\,\Big|_{\underline{r}=0} = \frac{\varrho}{2}\int \Phi_{ii}(\underline{k},t)\,d^3\underline{k} \qquad (4.15.15)$$

is the mean density of the kinetic energy in the fluid. Hence $\frac{\varrho}{2}V\,\Phi_{ii}(\underline{k},t)$ is the kinetic energy density in $\underline{k}$-space (V = vol

(*) In the homogeneous isotropic case the symmetrization symbols $\{\ \}$ in eqs. (4.14.3), (4.14.7) and (4.15.4) can be omitted. Why ?

ume of the fluid). Substituting (4.15.14) in (4.15.15) and performing the angular integrations we get

$$(4.15.16) \qquad V e_{kin}(t) = \varrho V \int_0^\infty E(k,t)\, dk .$$

So except for the constant factor ϱV the function $E(k,t)$ is the kinetic energy per unit interval of the wave number k. Of course, this result expresses the Wiener-Khinchin theorem in our present situation. It also explains the use of the letter E for the spectral density.

We return to eq. (4.15.13) which, as we have recognized, expresses the balance of the kinetic energy in the (one-dimensional) k-space. It is a striking feature of the present theory that a law of great importance, namely, the energy balance law, is obtained by formulating the equation for the 2nd-order correlation function. We considerate this a success even though the equation contains the unknown Fourier transform Υ of the 3rd-order correlation S.

One rigorous result can be derived from eq. (4.15.13) even though we do not know Υ. In fact, it follows from (4.15.5) and (4.15.12) that

$$(4.15.17) \qquad \int_0^\infty k^6\, \Upsilon(k,t)\, dk = 0$$

(prove this!). Hence the integration of eq. (4.15.13) gives

$$(4.15.18) \qquad \frac{\partial}{\partial t} \int_0^\infty E\,(k,t)\, dk + 2\nu \int_0^\infty k^2 E\,(k,t)\, dk = 0$$

The physical interpretation of this result is that the total kinetic energy of the fluid decreases to the same degree as the energy is dissipated by the viscous motion. In a way, this result can also be considered as a demonstration of correctness of our approach, since the physical model assumed here does not allow (total) energy changes other than those occurring in eq. (4.15.18).

On the other hand the right side of eq. (4.15.13) does not vanish. Evidently this means that there does not exist a law of detailed energy balance according to which the change of kinetic energy of waves (elements of turbulence) of a definite wave number k would be equal to the energy dissipated by these waves. This absence of a law of detailed energy balance clearly indicates that energy can be transferred from larger to smaller turbulence elements or vice versa.

The special form of the dissipated energy in (4.15.18) means that waves with large k (small λ) contribute more than waves with small k (large λ). The physical observation shows that the large elements decay into small elements by which then most of the energy is dissipated. This finding confirms the idea that energy flows from larger to smaller elements but not vice versa. On the other hand, the velocity of the particles is larger in the larger eddies so that one can assume that the kinetic energy is localized mainly in the large elements; at least, as long as these are present in a sufficient number.

c) Heisenberg's method

This approach is similar to a theory of von Weizsäcker which was established about the same time. The problem is to solve eq. (4.15.13) under a reasonable physical assumption about the unknown term

(4.15.19) $$F(k) \equiv 4\pi k^6 \Upsilon(k).$$

Heisenberg's assumption is in general agreement with the above discussion of the physical situation treated in this section. The main idea is that in a certain stage of the turbulence the smaller eddies - say those above a certain wave number K - are mutually in equilibrium in such a way that each one obtains just as much energy from larger eddies as it dissipates in the viscous motion. This means that the kinetic energy of all eddies above the wave number K remains constant. Everything else will follow from this assmption in connection with the basic equation (4.15.13).

We integrate (4.15.13) from 0 to K:

(4.15.20) $$\frac{\partial}{\partial t}\int_0^K E(k,t)\,dk = -2\nu\int_0^K k^2 E(k,t)\,dk + \int_0^K F(k,t)\,dk$$

This equation states that the kinetic energy of the large eddies $(k<K)$ decreases because they imply viscous flow in which energy is dissipated. In addition, there is just as much energy transferred to the small eddies $(k>K)$ as is dissipated by them. Hence the last integral is also equal to

$$-\nu\int_K^{\infty} k^2 E(k,t)dk \equiv \eta(K,t)\int_0^K k^2 E(k,t)dk. \quad (4.15.21)$$

Here we have introduced the so-called turbulence viscosity $\eta(K,t)$. Observe that the ratio of the two viscosities is

$$\frac{\eta(K,t)}{\nu} = \int_K^{\infty} k^2 E(k,t)dk \Big/ \int_0^K k^2 E(k,t)dk. \quad (4.15.22)$$

It equals the ratio of the energies dissipated by the small eddies and by the large eddies. Eq. (4.15.20) now becomes

$$\frac{\partial}{\partial t}\int_0^K E(k,t)dk = -2(\nu+\eta(K,t))\int_0^K k^2 E(k,t)dk. \quad (4.15.23)$$

For a fixed time t_0, say, we rewrite this equation in the form

$$\epsilon = 2(\nu+\eta(K))\int_0^K k^2 E(k)dk \quad (4.15.24)$$

and ϵ is the rate of decrease of the kinetic energy at t_0. If we write

$$\eta(K) \equiv \int_K^{\infty} \zeta(k)dk\,, \qquad \zeta(k) = \alpha\sqrt{E(k)/k^3} \quad (4.15.25)$$

then the turbulence viscosity appears as composed of the contributions $\zeta(k)$ of the elements with wave number $k > K$. α is dimensionless. Hence it can depend on k only via the dimensionless combination

$$\beta(k) \equiv k/(\epsilon/\nu^3)^{1/4}. \quad (4.15.26)$$

If we now make the physically plausible assumption that the internal friction depends strongly on the kinetic energy, but only weakly on it's rate of decrease, then we can rewrite eq. (5.15.24) in the form

$$\epsilon = 2\left(\nu + \alpha \int_K^\infty \sqrt{\frac{E(k)}{k^3}}\, dk\right) \int_0^K k^2 E(k)\, dk\,. \tag{4.15.27}$$

By solving this equation we obtain the spectrum of the kinetic energy density as

$$E(k) = C_1 \epsilon^{2/3} k^{-5/3} \qquad \beta(K) \ll 1 \tag{4.15.28}$$

$$E(k) = C_2 (\epsilon^2/\nu^4) k^{-7} \qquad \beta(K) \gg 1\,. \tag{4.15.29}$$

C_1 and C_2 are constants. It can be shown that $\beta(K)$ in these inequalities can be replaced by $\beta(k)$.

The result (4.15.28) was already obtained by Kolmogorov in 1941 using a somewhat different assumption about $F(k)$. A comparison with experiment mainly confirms eq. (4.15.28). The reader is referred to the discussion of Beran [B] , where also the references to experimental work are found.

It was our intention to present one of the impressive successes of statistical continuum mechanics. This investigation at the same time gave an important insight into the general methods applied in statistical continuum theories. It also showed that not everything can be achieved by only applying a certain formalism. In fact, it was the physical insight that led the workers in this field to the final success.

The turbulence viscosity is easily found from the calculated $E(k)$. It allows a good estimate of the increase of resistance against flow which is to be expected if laminar flow goes over into turbulent flow. This is another important result beside the energy spectrum.

As gratifying as these results may be, one has to be aware that only a modest aspect of the complex phenomenon of turbulence has been treated with success. There is a great open field left for further application of the statistical continuum theory. Unfortunately, the matter becomes more involved as soon as one has to go beyond the 2nd-order correlations.

CHAPTER V

Statistical theory of elastic materials with heterogeneous constitution

5.16. Introduction and statement of the physical problem

Many materials which appear to be homogeneous on a macroscopic scale turn out to be of a heterogeneous constitution on a much smaller scale to which we refer as the microscopic scale. Polycristalline aggregates as well as mixtures of several phases or materials belong to materials of this kind. These so-called composites have found increasing attention in recent time, both from the experimental and from the theoretical side.

The typical problem has been stated by W. Voigt as early as 1887. Let there be given a polycrystalline aggregate with known elastic moduli of the single constituents (grains) in a macroscopically homogeneous condition. Assume that the crystal orientations are distributed at random. Can the body be described macroscopically by effective elastic moduli in macroscopic applications and how are these moduli connected with the moduli of the single crystals?

Voigt has already given an approximate solution of this problem. Let

$$\underline{\sigma} = \underline{c}\,\underline{\varepsilon} \tag{5.16.1}$$

be the (anisotropic) Hooke's law of the single crystal in symbolic notation. In (5.16.1) $\underline{\sigma}$, $\underline{\varepsilon}$ and $\underline{c}$ denote the (symmetric) stress and strain tensors and the 4th-rank tensor of the elastic moduli of the single crystals respectively. $\underline{c}$ depends on the orientation of the particular crystal. Voigt makes the assumption that the strain is constant throughout the aggregate. In this case the volume average of eq. (5.16.1) can be written as

$$<\underline{\sigma}> = <\underline{c}\,\underline{\varepsilon}> = <\underline{c}><\underline{\varepsilon}> \tag{5.16.2}$$

If, on the other hand, we write the effective Hooke's law of the aggregate in the form

$$\underline{\Sigma} = \underline{C}\,\underline{\epsilon} \tag{5.16.3}$$

and identify the effective stress $\underline{\Sigma}$ and strain $\underline{\epsilon}$ with the volume average stress $<\underline{\sigma}>$ and strain $<\underline{\varepsilon}>$ respectively, then we see that in Voigt's approximation the effective modulus tensor becomes

$$\underline{C}^{V} = <\underline{c}>. \tag{5.16.4}$$

Many years later (1929) Reuss proposed to average Hooke's law in the inverse form assuming the stress to be constant throughout the body. In this case we have

(5.16.5) $$<\underline{\varepsilon}> = <\underline{C}^{-1}\underline{\sigma}> = <\underline{C}^{-1}><\underline{\sigma}> ,$$

and inverting this formula and comparing with eq. (5.16.3) we obtain in the Reuss approximation

(5.16.6) $$\underline{C}^R = <\underline{C}^{-1}>^{-1} .$$

Again many years later (1952) R. Hill has proved that under the assumption

(5.16.7) $$<\underline{\sigma}\,\underline{\varepsilon}> = <\underline{\sigma}><\underline{\varepsilon}>$$

the Voigt and Reuss approximations provide upper and lower bounds of the effective moduli, that is:

(5.16.8) $$C^R \leq C \leq C^V$$

where the inequalities refer to the eigenvalues of the Voigt matrices C_{ij} , C^R_{ij} , C^V_{ij} rather than to the components of the tensors. As we shall see later, eq. (5.16.7) is the condition under which the aggregate can at all be described by effective elastic moduli.

The investigation of Hill implies: If the whole information consists in the fact that the aggregate is built up from copper single crystals, for instance, and is macroscopically isotropic, then a more definite statement than eq. (5.16.8) is impossible. In order to get the bounds closer more information about the distribution of the grains is indispensable.

Since Hashin and Shtrikman (1962) have given closer bounds for the effective elastic moduli we conclude that they have inferred more information than that just mentioned. The Hashin-Shtrikman theory is only vaguely related to the theory to be developed in this chapter. We shall point out the connection at the proper moment.

Eq. (5.16.7) implies that the stress and strain in the aggregate are uncorrelated. It is hard to believe that eq. (5.16.7) should be valid in all cases. Hence a statement of the conditions under which the Hill condition is true deserves particular attention. This problem is treated in section 5.17. There we shall show that eq. (5.16.7) applies if, the body is infinite and if the stress state is produced only by forces of finite density actingon the (infinitely remote) surface. In this somewhat special situation does the concept of effective elastic moduli make sense.

Many attempts have been made in the past decades to improve the approximations of Voigt and Reuss. The arithmetic mean of $\underline{C}^V$ and $\underline{C}^R$ has been close to experimental results in many cases, although it lacks a physical basis. More physics is contained in the so-called self-consistent model of Hershey (1954) and Kröner (1958) which states that the effective elastic moduli should be equal to the moduli obtained by averaging the stress and strain in a spherical grain over all orientations. These quantities are calculated in this theory by solving the boundary value problem for a spherical elastic inclusion in an otherwise homogeneous matrix. Although the comparison with experiment was very favourable in many cases, it is clear from theoretical grounds that also the self-consistent effective moduli are not rigorous quantities. We shall explain by our rigorous statistical theory explained in section 5.18 why the Hill average usually denoted by $\underline{C}^{VRH}$ (VRH: Voigt-Reuss-Hill) and the self-consistent moduli $\underline{C}^{sc}$ are often so close to the true moduli.

The statistical theory of section 5.18 makes use of correlation functions up to infinite order. The method adopted is that of the Green's function; however, in a form which deviates slightly from the method explained in section 3.13. As in our outline of the statistical theory of turbulence we shall assume stationarity in space and macroscopic isotropy in a later stage. In addition we introduce the concept of perfect disorder which allows us to write down correlation functions up to infinite order as a kind of delta functions. The simple form of these functions makes it possible to perform the necessary integrations so that a final solution can be given in closed form.

The non-homogeneous case will be treated in section 5.19 whereas 5.10 contains recent results added in proof.

The never statistical theory of heterogeneous elastic materials was initiated independently by Beran and Molyneux, by Volkov and Klinskikh and by Lomakin obout 1965. Many ideas of these authors are incorporated in this chapter (*).

5.17. The Hill condition. Validity of the concept of effective elastic moduli

a) Foreword

In this section we specify the circumstances under which a linearly elastic heterogeneous material obeys an effective Hooke's law. The assumption that the body be a macroscopically homogeneous state will be introduced together with the ergodic theorem.

Let

$$\varepsilon_{ij} = s_{ijkl}\,\sigma_{kl} \tag{5.17.1}$$

be the inverse of the local Hooke's law in cartesian coordinates where s_{ijkl} is the local 4th-rank tensor of the elastic compliances. Effective elastic compliances S_{ijkl} can, then, be defined essentially in two ways, namely either by

$$\bar{\sigma}_{ij}\,\bar{\varepsilon}_{ij} = \bar{\sigma}_{ij}\,S_{ijkl}\,\bar{\sigma}_{kl} \tag{5.17.2}$$

(which means $\overline{s_{ijkl}\sigma_{kl}} = S_{ijkl}\,\bar{\sigma}_{kl}$) or via the energies

$$\overline{\sigma_{ij}\,\varepsilon_{ij}} = \bar{\sigma}_{ij}\,S_{ijkl}\,\bar{\sigma}_{kl} \tag{5.17.3}$$

where the bars indicate ensemble averages which we suppose to be equal to volume averages. Thus we assume the validity of the ergodic theorem. We prefer to speak in terms of ensembe means in

---

(*) A rather successful statistical theory of I.M. Lifshitz and L. N. Rosentsveig (1946) apparently did not attract much attention. Cf. the footnote in section 5.20, § b.

view of the theory of macroscopically non-homogeneous media to be developed in section 5.19 and also because the correlation functions introduced later are defined primarily as ensemble averages. In a way eq. (5.17.4) below is then a restatement of the Hill condition.

The definitions (5.17.2), (5.17.3) have been widely discussed in the literature. The condition that they be consistent is already quoted Hill condition, now written with subscripts,

(5.17.4) $$\overline{\sigma_{ij}\,\varepsilon_{ij}} = \bar{\sigma}_{ij}\,\bar{\varepsilon}_{ij}$$

This condition determines to which extent materials of a heterogeneous constitution follow the elasticity theory developed for materially uniform bodies. In fact, it is the condition for the validity of an effective Hoobe's law.

b) The Hill Condition

We first consider the case in which no self-stresses exist. The elastic strain can then be expressed by an elastic displacement field $u_j(\underline{x})$ according to

(5.17.5) $$\varepsilon_{ij} = (\partial_i u_j + \partial_j u_i)/2$$

where $\partial_i = \partial/\partial x_i$. The displacements follow from the densities $F_k(\underline{x})$ of volume forces and $A_k(\underline{x})$ of surface forces in the form (*)

$$u_j(\underline{x}) = \int_S G_{jk}(\underline{x},\underline{x}')A_k(\underline{x}')dS' + \int_V G_{jk}(\underline{x},\underline{x}')F_k(\underline{x}')dV'$$
(5.17.6)

(*) *Note that the tensor function $\underline{G}$ is self-adjoint, i.e.*
$G^{+}_{jk}(\underline{x},\underline{x}') = G_{kj}(\underline{x}',\underline{x}) = G_{jk}(\underline{x},\underline{x}')$.

where V, S are volume and surface, respectively, of the body. G_{jk} is the (fluctuating) Green's tensor of an elastic medium with variable elastic moduli c_{ijkl} . (Surface forces given).

The stresses obey the following equilibrium and boundary conditions

$$\partial_l \sigma_{kl} = -F_k, \quad n_l \sigma_{kl} = A_k \tag{5.17.7}$$

where n_l denotes the outward unit vector normal to the surface. Eq. (5.17.6) follows from eqs. (5.17.1, 5.17.5, 5.17.7).

We decompose σ_{kl} into its mean $\bar{\sigma}_{kl}$ and the deviation $\sigma'_{kl}(\underline{x})$ from the mean

$$\sigma_{kl}(\underline{x}) = \bar{\sigma}_{kl} + \sigma'_{kl}(\underline{x}), \quad \bar{\sigma}'_{kl} = 0 \tag{5.17.8}$$

and also define force densities $\bar{F}_k, F'_k, \bar{A}_k, A'_k$ by

$$\begin{aligned} \partial_l \bar{\sigma}_{kl} &\equiv -\bar{F}_k\,(=0), \quad \partial_l \sigma'_{kl} \equiv -F'_k, \\ n_l \bar{\sigma}_{kl} &\equiv \bar{A}_k, \quad n_l \sigma'_{kl} \equiv A'_k. \end{aligned} \tag{5.17.9}$$

With (5.17.7) and (5.17.9) we can rewrite eq. (5.17.6) in the form

$$\begin{aligned} u_j(\underline{x}) = v_j(\underline{x}) &+ \int_S dS' G_{jk}(\underline{x},\underline{x}') A'_k(\underline{x}') + \\ &+ \int_V dV' G_{jk}(\underline{x},\underline{x}') F'_k(\underline{x}') \end{aligned} \tag{5.17.10}$$

$$\begin{aligned} v_j(\underline{x}) &\equiv \int_S dS' G_{jk}(\underline{x},\underline{x}') \bar{A}_k = \\ &= \int dS'_l\, G_{jk}(\underline{x},\underline{x}')\, \bar{\sigma}_{kl} \end{aligned} \tag{5.17.11}$$

where $dS_l = n_l\, dS$.

We now use eqs. (5.17.10, 5.17.11) in order to calculate the expressions $\bar{\sigma}_{ij}\,\bar{\varepsilon}_{ij}$ and $\overline{\sigma_{ij}\,\varepsilon_{ij}}$ required in eqs. (5.17.2 - 5.17.3) and easily obtain

$$\bar{\sigma}_{ij}\,\bar{\varepsilon}_{ij} = \frac{1}{V}\int_S\int_S dS_i\, dS'_\ell\, G_{jk}(\underline{x},\underline{x}')\bar{\sigma}_{ij}\bar{\sigma}_{k\ell} + \tag{5.17.12}$$
$$+ \frac{1}{V}\int_S dS\, v'_k(\underline{x})\, A'_k(\underline{x}) + \frac{1}{V}\int dV\, v'_k(\underline{x})\, F'_k(\underline{x})$$

$$\overline{\sigma_{ij}\,\varepsilon_{ij}} = \bar{\sigma}_{ij}\,\bar{\varepsilon}_{ij} + \frac{1}{V}\int dS\, u'_k(\underline{x}) A'_k(\underline{x}) + \frac{1}{V}\int dV u'_k(\underline{x}) F'_k(\underline{x}) \tag{5.17.13}$$

where v'_k and u'_k are the deviations from the displacements v_k and u_k respectively explained in eqs. (5.17.6) and (5.17.11):

$$v'_k(\underline{x}) \equiv v_k(\underline{x}) - \bar{\alpha}_{k\ell}\, x_\ell \tag{5.17.14}$$
$$u'_k(\underline{x}) \equiv u_k(\underline{x}) - \bar{\varepsilon}_{k\ell}\, x_\ell$$

and

$$\alpha_{k\ell} = (\partial_k v_\ell + \partial_\ell v_k)/2\,. \tag{5.17.15}$$

In the derivation of (5.17.12, 5.17.13) we have also used the fact that

$$\frac{1}{V}\int_S dS\, x_\ell\, A'_k + \frac{1}{V}\int_V dV\, x_\ell\, F'_k = \overline{\sigma'_{k\ell}} = 0\,. \tag{5.17.16}$$

In particular, we see from eq. (5.17.13) that the Hill condition is satisfied if and only if

$$\frac{1}{V}\int_S dS\, u'_k(\underline{x})\, A'_k(\underline{x}) + \frac{1}{V}\int_V dV\, u'_k(\underline{x})\, F'_k(\underline{x}) = 0 \tag{5.17.17}$$

which also means $\overline{\sigma'_{ij}\varepsilon'_{ij}} = 0$. This result is rather trivial and could have been obtained more directly from eq. (5.17.4). We have chosen the more elaborate way because it gives the additional important result (5.17.18) below.

So far everything has been exact.

c) The effective Hooke's law

In order to specify in some generality conditions under which a heterogeneous material obeys an effective Hooke's law we introduce the assumption that the body is in a macroscopically homogeneous state. Strictly speaking this limitation requires, at the same time, an infinite body. Hence we consider a large finite body which in the limit becomes infinite. We further assume that the local compliances as well as the surface and volume force densities are bounded. Then also u'_k and v'_k are bounded:

According to the laws of linear elasticity theory, a point-like force singularity acting in the interior or on the surface of the body causes a displacement field (relative to the rest of the body) which is singularly infinite at the point of action. Conversely, a displacement field which is singularly infinite at a point can only be produced by a point-like force singularity.

If external forces of an infinite density act on a small (volume or surface) region of the body then the resulting displacement field can be calculated as the superposition of displacements of point forces. Obviously, the displacements in the region are infinite relative to the rest of the body. Converse ly, if some region of the body is displaced infinitely relative to the rest of the body we expect that forces of an infinite density are acting in the region. If such forces are excluded from the body then the mentioned relative displacements cannot occur.

Displacement fluctuations are relative displacements with respect to a certain average configuration of the body. If they are infinite in some region then corresponding forces of infinite density must be present. If such forces are excluded, then the displacement fluctuations cannot be infinite. In other words, if A_k and F_k are bounded then also u_k' and v_k' are bounded.

In the limit $V \to \infty$ the surface integrals in (5.17.12) and (5.17.13) vanish when multiplied by $1/V$ because the integrands are bounded.

In the volume integral of (5.17.12) the fluctuating displacement v_k' is correlated to the distribution of the local elastic compliances via the Green's tensor. It follows that the volume integral does not vanish if and only if F_k' also is correlated to the compliances. Note that F_k' can be prescribed independently of v_k'. A corresponding statement holds for the surface integral in (5.17.12) if we consider a finite body (i.e. not a macroscopically homogeneous situation).

In the volume integral of eq. (5.17.13) u_k' is always correlated to F_k' if such volume forces are present at all. In that case this integral gives a contribution even in a homogeneous body. A similar remark holds for the surface integral in

(5.17.13) in the case of a finite body.

So we have obtained the result: A linearly elastic heterogeneous material in an ergodic situation obeys an effective Hooke's law for the average stress and strain with an effective tensor of elastic compliances which then have the form (cf. eqs. (5.17.3, 5.17.12))

$$S_{ijkl}=\frac{1}{V}\iint\limits_{S\,S} dS_i\, dS'_l\, G_{jk}(\underline{x},\underline{x}')_{\{ijkl\}}=\frac{1}{V}\iint\limits_{V\,V} dV dV'\,\partial_i \partial'_l G_{jk}(\underline{x},\underline{x}')_{\{ijkl\}} \qquad (5.17.18)$$

(i) if the local elastic compliances and surface force densities are bounded;

(ii) if the body is in a macroscopically homogeneous situation, hence infinite in the limit;

(iii) if no volume forces are present.

This theorem has been derived under the assumption that no internal stresses are present (*).

In eq. (5.17.18) the suffix $\{ijkl\}$ indicates symmetrization so that S_{ijkl} assumes the usual symmetry:

$$S_{ijkl} = S_{jikl} = S_{ijlk} = S_{klij} \, . \qquad (5.17.19)$$

The justification for this symmetrization follows from eq. (5.17.12).

In addition to the above result a particular case deserves mention. Eqs. (5.17.17, 5.17.18) are satisfied, too, if beside $F'_k = 0$ also $A'_k = 0$ holds in a finite body. However,

(*) Note added in proof: Meanwhile, necessary and sufficient conditions have been formulated, see section 5.20, § d.

we then do not have a true macroscopically homogeneous situation. Rather a boundary layer effect appears. In the extreme case of the perfectly disordered body, treated later, the boundary layer disappears and we have then, in fact, the important case where a finite body seems to obey an effective Hooke's law. In order to prove that it really does obey such a law one has to show that the surface integrals in (5.17.12, 5.17.13) vanish also if $A'_k \neq 0$. This can be done, see E. Kröner (1973). For further results cf. section 5.20.

d) Internal Stresses

The foregoing considerations can easily be extended to include the case when internal stresses are present. Let $\varepsilon^s_{ij}(\underline{x})$ be the spontaneous or self-strain which we consider as the source of the internal stresses. As is well-known, neither ε^s_{ij} nor the elastic strain ε_{ij} can be expressed in terms of a displacement field as was done in eq. (5.17.5). An equation of this type only exists for the total strain

(5.17.20) $$\varepsilon^T_{ij} = \varepsilon^s_{ij} + \varepsilon_{ij}$$

which has the form

(5.17.21) $$\varepsilon^T_{ij} = \left(\partial_i u^T_j + \partial_j u^T_i\right)/2$$

with u^T_j as total displacement.

We multiply (5.17.20) by the tensor $c_{ijkl} \equiv s^{-1}_{ijkl}$

of the local elastic moduli and obtain

$$\sigma^{T}_{k\ell} = \sigma^{S}_{k\ell} + \sigma_{k\ell} . \tag{5.17.22}$$

This means that we have defined two stress-type tensors $\sigma^{T}_{k\ell}$ and $\sigma^{S}_{k\ell}$ by

$$\varepsilon^{T}_{ij} = s_{ijk\ell}\sigma^{T}_{k\ell} , \quad \varepsilon^{S}_{ij} = s_{ijk\ell}\sigma^{S}_{k\ell} . \tag{5.17.23}$$

We substitute (5.17.22) in the equilibrium and boundary conditions (5.17.7) and obtain

$$\partial_\ell \sigma^{T}_{k\ell} = -F^{T}_{k} , \quad n_\ell \sigma^{T}_{k\ell} = A^{T}_{k} \tag{5.17.24}$$

where

$$F^{T}_{k} \equiv F^{S}_{k} + F_{k} , \quad A^{T}_{k} \equiv A^{S}_{k} + A_{k} \tag{5.17.25}$$

and

$$\begin{aligned} F^{S}_{k} &\equiv -\partial_\ell \sigma^{S}_{k\ell} = -\partial_\ell \left(c_{k\ell mn}\varepsilon^{S}_{mn} \right) \\ A^{S}_{k} &\equiv n_\ell \sigma^{S}_{k\ell} = n_\ell c_{k\ell mn}\varepsilon^{S}_{mn} \end{aligned} \tag{5.17.26}$$

are fictive volume and surface forces which replace the original stress sources ε^{S}_{mn}. This method is generally attributed to Duhamel and Neumann who treated the special case of thermal stresses.

Our former results, in particular the eqs. (5.17.12, 5.17.13), were derived essentially from eqs. (5.17.1, 5.17.5,

5.17.7). A comparison of these equations with the eqs. (5.17.23, 5.17.21, 5.17.24) indicates that the result of the present calculation will be equations of the type (5.17.12-5.17.16) where the "elastic" quantities are replaced by the "total" quantities. The conditions which ensure the validity of (5.17.18) are now complemented by the requirement that the internal stress sources are distributed in such a way that no fictive volume forces F_k^S are present.

Since the theory of internal stresses states that these vanish if the fictive volume forces disappear we have obtained that (5.17.18) is true when in addition to the conditions (i) - (iii) also the requirement that

(IV) no internal stresses are present,

is fulfilled.

e) Consequences

No real body is infinite. So we are concerned - as it is often the case in physics - with an idealized situation which does not really occur in nature. However, our result will be a good approximation to real situations, if the external dimensions of the composite are sufficiently large compared to the typical grain diameter.

An important case of volume forces is that of inertial forces. Clearly these forces when present will be correlated to the local compliances. In this case the Hill condition will not be satisfied and, therefore, the elastic compliances in this situation cannot be defined consistently. As a consequence the relationship between the elastic compliances defined by eq. (5.17.18) and the velocities of sound are more complex than in homogeneous bodies. In near future we hope to say more about this effect which is probably small though not beyond experimental accuracy.

Internal stresses too are usually correlated to the distribution of the local compliances. There are consequences even for the static measurement of elastic compliances of a body with internal stresses, and even when these stresses do not vary under the external loading. In fact, the two important integrals do not vanish in this case, i.e. Hill's condition is not satisfied and we have no consistently defined effective compliances which describe the response of the body in this experiment. The situation is then rather complex.

5.18. The rigorous (statistical) theory of the effective elastic moduli

a) Preliminaries

In the last section we have stated the conditions under which a linearly elastic heterogeneous material which is found to be in a macroscopically homogeneous state obeys an effective Hooke's law. The next task is then to calculate the effective elastic moduli or compliances from the local moduli or compliances of the constituents. This is clearly a statistical problem since we are dealing with the random distributions of the quantities concerned.

Let us define the 4th-rank tensor $\underline{C} \equiv (C_{ijkl})$ of the effective elastic moduli by the relation

$$(5.18.1) \qquad \bar{\sigma}_{ij} = \overline{c_{ijkl}\,\varepsilon_{kl}} = C_{ijkl}\,\bar{\varepsilon}_{kl}.$$

Resolution of c_{ijkl} and ε_{kl} into mean and deviation leads to

$$(5.18.2) \qquad C_{ijkl}\,\bar{\varepsilon}_{kl} = \bar{c}_{ijkl}\,\bar{\varepsilon}_{kl} + \overline{c'_{ijkl}\,\varepsilon'_{kl}}.$$

Since $\bar{c}_{ijkl}$ is easily obtained from c_{ijkl}, the main problem will be to calculate the mixed 2nd-order correlation function $\overline{c'_{ijkl}(\underline{x})\varepsilon'_{kl}(\underline{x})}$. For this purpose we need information on the distribution of the local elastic moduli. Of course, this can only be the macroscopic information since the details of the microscopic distribution are neither known nor of interest. We could give the macroscopic (or probabilistic) information by means of the probability density functional of c_{ijkl}. However, as in the theory of turbulence, the more successful approach utilizes correlation functions in c_{ijkl}, or, more conveniently, in c'_{ijkl}. We shall find that we need the infinite set of correlation functions of c'_{ijkl} in order to express $\overline{c'_{ijkl}(\underline{x})\,\varepsilon'_{kl}(\underline{x})}$ exactly. This remark shows that our problem is not at all trivial.

We shall apply the method of the Green's function as explained in section 3.13; however in a somewhat modified form.

b) The general theory

We start our calculation by writing down the deterministic equation governing the strain deviation $\varepsilon'_{k\ell}(\underline{x})$. This equation is the stress equilibrium equation with the local Hooke's law incorporated in it. Using σ'_{ij} rather than σ_{ij} we obtain

$$0 = \partial_j^{(1)} \left[c_{ijk\ell}(\underline{x}_1)\,\varepsilon_{k\ell}(\underline{x}_1)\right]' = \partial_j^{(1)} \left\{ \left[\bar{c}_{ijk\ell} + c'_{ijk\ell}(\underline{x}_1)\right] \left[\bar{\varepsilon}_{k\ell} + \varepsilon'_{k\ell}(\underline{x}_1)\right] \right\}' \qquad (5.18.3)$$

or

$$\bar{c}_{ijk\ell}\,\partial_j^{(1)} \varepsilon'_{k\ell}(\underline{x}_1) = -\left[\partial_j^{(1)} c'_{ijk\ell}(\underline{x}_1)\right] \bar{\varepsilon}_{k\ell} - \partial_j^{(1)} \left[c'_{ijk\ell}(\underline{x}_1)\,\varepsilon'_{k\ell}(x_1)\right]' . \qquad (5.18.4)$$

From this equation we could derive macroscopic equations for the correlation functions as described in sections 3.11, 3.12. We prefer the more direct solution of the problem in terms of the Green's tensor of elasticity theory. We can do this here because the theory is linear in contrast to the Navier-Stokes equations.

It is important to note that we shall use the Green's tensor of a fictive homogeneous body with elastic moduli $\bar{c}_{ijk\ell}$. This is the reason why we have written the equilibrium condition in the particular form (5.18.4). It is not necessary to proceed in this way, but we find it convenient. In fact, the Green's tensor of a homogeneous medium is not a fluctuating quantity. In this respect the present method deviates from that outlined in section 3.13.

We can immediately write down the Green's tensor in Fourier representation. It has the form (*)

$$G_{\ell m}(\underline{x}_{12}) = \frac{1}{8\pi^3}\int \Delta^{-1}_{\ell m}(i\underline{k})e^{i\underline{k}\cdot\underline{x}_{12}}d^3\underline{k} \qquad (5.18.5)$$

and satisfies the equation

$$\Delta_{j\ell}(\underline{\nabla})\, G_{\ell m}(\underline{x}_{12}) + \delta(\underline{x}_{12})\,\delta_{jm} = 0 \qquad (5.18.6)$$

where $\underline{x}_{12} \equiv \underline{x}_2 - \underline{x}_1$ and

$$\Delta_{j\ell}(\underline{\nabla}) \equiv \bar{c}_{ijk\ell}\, \partial^{(1)}_i \partial^{(1)}_k . \qquad (5.18.7)$$

$\delta(\underline{x}_{12})$ is the Dirac's delta function.

In the special case of an isotropic tensor $\bar{c}_{ijk\ell}$ with Lamé's constants $\bar{\lambda}$, $\bar{\mu}$ one finds

$$\Delta^{-1}_{\ell m}(i\underline{k}) = \frac{-(\bar{\lambda}+\bar{\mu})k_\ell k_m + (\bar{\lambda}+2\bar{\mu})k^2\delta_{\ell m}}{\bar{\mu}(\bar{\lambda}+2\bar{\mu})k^4} \qquad (5.18.8)$$

A particular solution of eq. (5.18.4) can be written as

$$\varepsilon'_{k\ell}(\underline{x}_1) = -\int dV_2\, G_{i(\ell,k)}(\underline{x}_{12})\Big[\big(\partial^{(2)}_j c'_{ijmn}(\underline{x}_2)\big)\bar{\varepsilon}_{mn} + \partial^{(2)}_j\big(c'_{ijmn}(\underline{x}_2)\varepsilon'_{mn}(x_2)\big)'\Big] \qquad (5.18.9)$$

which can be verified easily by insertion. Here the comma means partial derivation with respect to the argument and the paren-

---

(*) Observe that this is *<u>not</u>* *the Green's tensor used in section 17. Why not?*

theses around subscripts indicate symmetrization. This derivation and symmetrization imply that $\varepsilon'_{kl}(\underline{x}_1)$ has the form of a so-called deformator, i.e. satisfies the de St. Venant compatibility conditions. In eq. (5.18.9) as in all later equations the integrations in $\underline{x}$ are taken over the infinite space.

We can obtain further particular solutions of eq. (5.18.4) by adding solutions of eq. (5.18.4) with $\bar{\varepsilon}_{kl} = 0$. However, the investigation of section 5.17 has shown that such solu tions would imply $\bar{\sigma}_{ij} = 0$; hence they would not contribute to $\overline{c'_{ijkl}\varepsilon'_{kl}}$. This means that the solution (5.18.9) will fulfill the purpose (*).

This solution is of an implicit form. In fact, it is an integro-differential equation for $\varepsilon'_{kl}(\underline{x}_1)$, when $c'_{ijkl}(\underline{x}_1)$ and $\bar{\varepsilon}_{kl}$ are given. Eq. (5.18.9) can be solved by iteration if convergence is assumed. We set

$$\varepsilon'_{kl} = \overset{(1)}{\varepsilon'_{kl}} + \overset{(2)}{\varepsilon'_{kl}} + \overset{(3)}{\varepsilon'_{kl}} + \dots \qquad (5.18.10)$$

and easily obtain

$$\left\{\begin{aligned} \overset{(1)}{\varepsilon'_{kl}}(\underline{x}_1) &= -\int dV_2\, G_{i(k,l)}(\underline{x}_{12})\,\partial_j^{(2)} c'_{ijmn}(\underline{x}_2)\,\bar{\varepsilon}_{mn} \\ \overset{(2)}{\varepsilon'_{kl}}(\underline{x}_1) &= \int dV_2\, G_{i(k,l)}(\underline{x}_{12})\,\partial_j^{(2)}\Big[c'_{ijpq}(\underline{x}_2)\ dV_3\, G_{r(p,q)} \times \\ &\qquad \times (\underline{x}_{23})\,\partial_s^{(3)} c'_{rsmn}(\underline{x}_3)\,\bar{\varepsilon}_{mn}\Big]' \end{aligned}\right. \qquad (5.18.11)$$

etc.. For the later convenience we rewrite these equation in the form

---

(*) In my original work (E. Kröner, 1967) the last prime of eq. (5.18.9) was missing. As pointed to me by P. Dederichs and R. Zeller this prime ensures that the later used particular solution of eq. (5.18.9) has the mean zero, as it should be. Due to this missing prime my final solution of 1967 needs a fourth-order correction.

$$(5.18.12)\quad \begin{cases} \overset{(1)}{\varepsilon}{}'_{k\ell}(\underline{x}_1) = -\int dV_{12}\, G_{i(k,\ell)}(\underline{x}_{12})\, \partial_j^{(12)}\, c'_{ijmn}(\underline{x}_1+\underline{x}_{12})\,\bar{\varepsilon}_{mn} \\ \overset{(2)}{\varepsilon}{}'_{k\ell}(\underline{x}_1) = \int dV_{12}\, G_{i(k,\ell)}(\underline{x}_{12})\, \partial_j^{(12)} \int dV_{23}\, G_{r(p,q)}(\underline{x}_{23})\, \partial_s^{(23)} \times \\ \qquad \times \left[c'_{ijpq}(\underline{x}_1+\underline{x}_{12})\, c'_{rsmn}(\underline{x}_1+\underline{x}_{12}+\underline{x}_{23})\right]'\bar{\varepsilon}_{mn} \end{cases}$$

etc. where, for example, dV_{12} and $\partial_j^{(12)}$ are volume element and differentiation symbol in the space of $\underline{x}_{12}$. The independent variables in the integrals are $\underline{x}_1, \underline{x}_{12}, \underline{x}_{23}$ etc.

Just as in the turbulence problem we shall assume that averaging - which implies integration over $\underline{x}_1$ - commutes with integrations and differentiations referring to the variables $\underline{x}_{12}$, $\underline{x}_{23}$ etc. . Then we can, first of all, state that the solution (5.18.10, 5.18.12) satisfies $\overline{\varepsilon'_{k\ell}} = 0$ as it should do for the sake of consistency. (Prove this!)

Furthermore, we multiply the eqs. (5.18.12) by $c'_{abk\ell}(\underline{x}_1)$ and then take the average. In this way we obtain

$$(5.18.13)\quad \begin{cases} \overline{c'_{abk\ell}(\underline{x}_1)\,\overset{(1)}{\varepsilon}{}'_{k\ell}(\underline{x}_1)} = -\int dV_{12}\, G_{i(k,\ell)}(\underline{x}_{12})\, \partial_j^{(12)} \times \\ \qquad \times\, \overline{c'_{abk\ell}(\underline{x}_1)\, c'_{ijmn}(\underline{x}_2)}\,\bar{\varepsilon}_{mn} \\ \overline{c'_{abk\ell}(\underline{x}_1)\,\overset{(2)}{\varepsilon}{}'_{k\ell}(\underline{x}_1)} = \int dV_{12}\, G_{i(k,\ell)}(\underline{x}_{12})\, \partial_j^{(12)} \int dV_{23}\, G_{r(p,q)} \times \\ \qquad \times\, \partial_s^{(23)}\, \overline{c'_{abk\ell}(\underline{x}_1)\left[c'_{ijpq}(\underline{x}_2)\, c'_{rsmn}(\underline{x}_3)\right]'}\,\bar{\varepsilon}_{mn} \end{cases}$$

etc.. These equations together with (5.18.2) and (5.18.10) solve the task of expressing the effective tensor $\underline{C}$ in terms of the macroscopic information about the local tensor $\underline{c}$. Observe that the Green's tensor $\underline{G}$ is completely determined by the local $\underline{c}$ according to eqs. (5.18.5) and (5.18.7). The solution gives $\underline{C}$ in terms of the infinite set of correlation functions of $\underline{c}(\underline{x})$.

Except for the conditions derived in the last section no assumptions have been made about the distribution of the local elastic moduli. Hence the equations we have obtained apply to polycrystalline aggregates as well as to any other composite materials and even in macroscopically anisotropic situations. Of course, macroscopic homogeneity was assumed.

We now summarize the results obtained so far in an abridged notation which explains itself by comparison with the former equations. By partial integration in which the surface integrals do not contribute we obtain eq. (5.18.9) in the new form

$$\underline{\varepsilon}' = -\underline{\Gamma}\,\underline{c}'\bar{\underline{\varepsilon}} - \underline{\Gamma}P\underline{c}'\underline{\varepsilon}' \equiv -\underline{\Gamma}\,P\underline{c}'\underline{\varepsilon} \qquad (5.18.14)$$

where P is Beran and McCoy's "prime operator" acting on the full factor to its right hand side so that, for instance, $Pa(b-c) \equiv [a(b-c)]'$. $\underline{\Gamma}$ is the modified Green's tensor with the components. (*)

(*) Note that the tensor function $\underline{\Gamma}$ is self-adjoint because $\underline{G}$ has this property: $\Gamma^{+}_{ijkl}(\underline{x}_1,\underline{x}_2) \equiv \Gamma_{klij}(\underline{x}_2,\underline{x}_1) = \Gamma_{ijkl}(\underline{x}_1,\underline{x}_2)$. This is true also for those tensors which are formed with Green's tensors $\underline{G}$ obeying special boundary conditions.

(5.18.15) $$\Gamma_{ijkl}(\underline{x}_1, \underline{x}_2) = \partial_k^{(1)} \partial_j^{(2)} G_{il}(\underline{x}_1, \underline{x}_2)\Big|_{(ij),(kl)}$$

where the suffices (ij), (kl), denote symmetrization in the indicated subscripts, respectively. (Convince yourself that the signs in (5.18.14) are correct.)

Incidentally, the modified Green's tensor relates strain and stress in the form

(5.18.16) $$\varepsilon_{ij}(\underline{x}_1) = \int_V \Gamma_{ijkl}(\underline{x}_1, \underline{x}_2)\,\sigma_{kl}(\underline{x}_2)\,dV_2 \quad (\underline{\varepsilon} = \underline{\Gamma}\,\underline{\sigma}).$$

This equation even applies to a finite medium if the Green's tensor $\underline{G}$ of the infinite medium is replaced by the Green's tensor $\underline{G}^{(2)}$ of the 2nd boundary value problem of the finite medium. $\underline{G}$ and the corresponding $\underline{\Gamma}$ depend on $\underline{x}_1$ and $\underline{x}_2$ separately, i.e. not only on the combination $\underline{x}_1 - \underline{x}_2$.

After multiplication with $\underline{c}'$ from the left we can rewrite eq. (5.18.14) in the form

(5.18.17) $$(\underline{I} + \underline{c}'\,\underline{\Gamma}\,P)\,\underline{c}'\,\underline{\varepsilon}' = -\,\underline{c}'\,\underline{\Gamma}\,\underline{c}'\,\bar{\underline{\varepsilon}}$$

where

$$\underline{I} = (I_{ijkl}) = \delta(\underline{x}_1, \underline{x}_2)\,(\delta_{ik}\delta_{jl} + \delta_{il}\delta_{jk})/2$$

is the (4th rank) unit tensor of our function space. It follows formally that

$$\underline{c}'\underline{\varepsilon}' = -\left(\underline{I} + \underline{c}'\underline{\Gamma}P\right)^{-1}\underline{c}'\underline{\Gamma}\underline{c}'\bar{\underline{\varepsilon}} . \qquad (5.18.18)$$

A comaprison with eq. (5.18.2) leads to

$$\underline{C} = \bar{\underline{c}} - \overline{\left(\underline{I} + \underline{c}'\underline{\Gamma}P\right)^{-1}\underline{c}'\underline{\Gamma}\underline{c}'} . \qquad (5.18.19)$$

Formal development of this symbolical equation into a geometric series gives

$$\underline{C} = \bar{\underline{c}} - \overline{\underline{c}'\underline{\Gamma}P\underline{c}'} + \overline{\underline{c}'\underline{\Gamma}P\underline{c}'\underline{\Gamma}P\underline{c}'} - \overline{\underline{c}'\underline{\Gamma}P\underline{c}'\underline{\Gamma}\underline{c}'\underline{\Gamma}P\underline{c}'} + \dots \qquad (5.18.20)$$

where we have used $P\underline{c} = P\underline{c}'$. It is easily shown that this equation is equivalent to the combination of eqs. (5.18.2), (5.18.10) and (5.18.13). Hence it is again the solution of our problem, namely to express $\underline{C}$ in terms of the correlation functions of $\underline{c}$. Note that the Γ's do not fluctuate and, therefore can be put outside of the averages.

Although the solution (5.18.19) looks rather simple it is extremely complex in reality since the terms of the series (5.18.20) are multiple integrals of increasing order. Series of this kind occur in various fields of physics, for instance in the general scattering theory where they are called cluster developments and illustrated by graphs.

If in eq. (5.18.20) the P's are omitted the result of my 1967-paper remains. Apparently, this result is correct to the 3rd order correlations (show this!).

The eqs. (5.18.20) imply that it is not necessary to have the complete, i.e. microscopic, information about the moduli distribution in a specimen to get its correct tensor C. It suffices to know all the correlation functions of $\underline{c}$. Many specimens with different microscopic details of $\underline{c}(\underline{x})$ can have identical correlation functions. All these specimens have the same expectation of their effective elastic moduli. To give the information about the material in the form of correlation functions is a quite natural way. Often the whole set of correlation functions may not be known but only a few of them. In such a case previous investigations of Beran and Molyneux lead to the expectation that our equations give an upper bound for the moduli tensor $\underline{\mathsf{C}}$ when all unknown correlation functions are set equal to zero (*).

The work of Beran and Molyneux also contains the result that information given in terms of the local compliances s_{ijkl} rather than in the local moduli can lead to lower bounds. The Voigt and Reuss bounds which are calculated from the 1st-order correlation functions (= means) of $\underline{c}$ and $\underline{s}$ respectively, and the Hashin-Shtrikman bounds which follow from additional information essentially that of the 2nd-order correlations of $\underline{c}$ and $\underline{s}$ belong to this general scheme. For details we refer to the

(*) Note added in proof : Meanwhile, this has been proved by P. Dederichs and R. Zeller (1972) See also section 5.20.

work of Beran and Molyneux and of Dederichs and Zeller (1972).

c) <u>The perfectly disordered polycrystalline aggregate</u>

It will often be the case that information about the heterogeneous constitution of a body is not given in terms of correlation functions but by some other quantities. In order to apply our theory, one then has to calculate the correlation functions from the given information.

There are two extreme cases in which all the correlation functions can be written down immediately. These are the cases of perfect order and of perfect disorder. The first case does not fit well into a statistical theory and is therefore not treated here.

Perfect disorder is defined as the situation in which the moduli distribution is statistically independent over all points. This means that nothing particular can be predicted for the elastic moduli at point $\underline{x}_1$ if the moduli are known at point $\underline{x}_2$, or at points $\underline{x}_2, \underline{x}_3$, or at points $\underline{x}_2, \underline{x}_3, \underline{x}_4$ etc.. In a strict sense such a situation is not realistic. In fact, if point $\underline{x}_2$ is very close to point $\underline{x}_1$, then there is a high probability that $\underline{x}_2$ is within the same grain as is $\underline{x}_1$. In that case an information about $\underline{x}_2$ does mean something for the prediction at $\underline{x}_1$. If, on the other hand, the separation of the two points is large enough then there is a very high probability that the two points are in different grains in which a result at $\underline{x}_2$ does not influence the expectation at $\underline{x}_1$. Large enough means a few mean grain diameters.

The foregoing considerations show that perfect disorder (i.e. the limiting case for the grain sizes tending to zero) is an unrealistic limiting case. However, if the dimensions of the body are very large compared to the grain sizes - in other words, if the number of grains is very large - then the perfect disorder limiting case will be an extremely good approximation to the real body. This observation justifies that we consider this case in more detail.

An extension of the investigation leading to the result (see section 2.6) that the correlation function $\overline{u'(\underline{x}_1)u'(\underline{x}_2)} = 0$ for $\underline{x}_1 \neq \underline{x}_2$ if the values of the random variable u at points $\underline{x}_1$ and $\underline{x}_2$ are independent shows us that the nth-order correlation functions of c'_{ijkl} vanish when at least one single point exists in the considered set $\underline{x}_1, \underline{x}_2 \ldots \underline{x}_n$ which is separated from all other points. Hence the nth-order correlation functions can be non-zero only when either all points $\underline{x}_1$, $\underline{x}_2 \ldots \underline{x}_n$ coincide or when the set can be divided into subsets of at least two points each where all points in a subset coincide.

Because of the mathematical complexity the contribution of the latter correlation functions has not yet been calculated and will, therefore, be neglected in our treatment of the perfectly disordered material. Since the neglected functions appear only in terms of order higher than 3 it is justified to omit also the complication implied by the P-operations in eq. (5.18.20). As mentioned, these too affect only the terms of order higher then 3. Eq. (5.18.20) then becomes

$$\underline{C} = \underline{\bar{c}} - \overline{\underline{c}'\underline{\Gamma}\,\underline{c}'} + \overline{\underline{c}'\underline{\Gamma}\,\underline{c}'\underline{\Gamma}\,\underline{c}'} - \ldots \qquad (5.18.21)$$

where all deviations $\underline{c}'$ refer to the same point $\underline{x}_1$. The correlation functions occurring in (5.18.21) are of all orders n and have the form

$$\overline{\underline{c}'(\underline{x}_1)\,\underline{c}'(\underline{x}_2)\ldots c'(\underline{x}_n)} = \overline{\underline{c}'(\underline{x}_1)\underline{c}'(\underline{x}_1)\ldots\underline{c}'(\underline{x}_1)}\,\delta_{\underline{x}_{12}}\delta_{\underline{x}_{23}}\ldots\delta_{\underline{x}_{n-1,n}}. \qquad (5.18.22)$$

Here $\delta_{\underline{x}_{12}}$ is a delta type function defined by

$$\delta_{\underline{x}_{12}} = \begin{matrix} 1 \\ 0 \end{matrix} \quad \text{for} \quad \begin{matrix} \underline{x}_1 = \underline{x}_2 \\ \underline{x}_1 \neq \underline{x}_2 \end{matrix} \qquad (5.18.23)$$

Since $\overline{c'(\underline{x}_1)\underline{c}'(\underline{x}_1)\ldots\underline{c}'(\underline{x}_1)}$ is a constant in a macroscopically homogeneous material we can set it outside the integrals in (5.18.13). Thus the integrals are factorized and we can write, for example

$$\overline{c'_{abkl}(\underline{x}_1)\varepsilon_{kl}(\underline{x}_1)} = \overline{c'_{abkl}(\underline{x}_1)\,c'_{ijpq}(\underline{x}_1)\,c'_{rsmn}(\underline{x}_1)}\;E_{klij}E_{pqrs}\bar{\varepsilon}_{mn} \qquad (5.18.24)$$

where E_{ijkl} has the symmetry of the tensor $\bar{c}_{ijkl}$ and is defined by

$$E_{ijkl} = \int dV_{12}\,\Gamma_{ijkl}(\underline{x}_{12})\,\delta_{\underline{x}_{12}}. \qquad (5.18.25)$$

This tensor has been used by Kneer (1965) in connection with the anisotropic inclusion problem. The values of the components of $\underline{E}$ were given explicitly in terms of $\bar{\underline{c}}$ for cubic and hexagonal symmetry. For isotropic $\bar{\underline{c}}$ (no texture) the integral (5.18.25) can be evaluated with the help of (5.18.8). One finds that $\underline{E}$ is an isotropic tensor which is completly determined, for example, by

$$E_{1111} + 2E_{1122} = \frac{1}{3\bar{k}+4\bar{\mu}} ,$$

(5.18.26)
$$2E_{1212} = E_{1111} - E_{1122} = \frac{3\bar{k}+6\bar{\mu}}{5\bar{\mu}(3\bar{k}+4\bar{\mu})}$$

where $\bar{k}$ and $\bar{\mu}$ are the compression and shear modulus of the tensor $\bar{\underline{c}}$ (see below).

Using the tensor $\underline{E}$, now given, we can write the expression $\overline{\underline{c}'\underline{\varepsilon}'}$ in the condensed form

(5.18.27)
$$\overline{\underline{c}'\underline{\varepsilon}'} = \left(-\overline{\underline{c}'\underline{E}\,\underline{c}'} + \overline{\underline{c}'\underline{E}\,\underline{c}'\underline{E}\,\underline{c}'} - \overline{\underline{c}'\underline{E}\,\underline{c}'\underline{E}\,\underline{c}'\underline{E}\,\underline{c}'} + \dots\right)\bar{\underline{\varepsilon}} \equiv$$
$$\equiv -\overline{\left(\frac{\underline{c}'\underline{E}\,\underline{c}'}{\underline{I}+\underline{E}\,\underline{c}'}\right)}\,\bar{\underline{\varepsilon}}$$

where $\underline{I} = (I_{ijkl}) = (\delta_{ik}\delta_{jl} - \delta_{il}\delta_{jk})/2$. and the integrations have disappeared. The effective elastic moduli tensor follows from (5.18.27) together with (5.18.2):

(5.18.28)
$$\underline{C} - \bar{\underline{c}} = -\overline{\left(\frac{\underline{c}'\underline{E}\,\underline{c}'}{\underline{I}+\underline{E}\,\underline{c}'}\right)} \qquad (*)$$

(*) These results were obtained independently by V.V. Bolotin and V.N. Moskalenko (1968/69).

The tensor $\bar{\underline{c}}$ is easily calculated from $\underline{c}$ which also gives $\underline{c}' = \underline{c} - \bar{\underline{c}}$. $\underline{E}$ was already expressed in terms of the components of $\bar{\underline{c}}$. Hence the right hand side of (5.18.28) is practically given in terms of the tensor $\underline{c}$ alone. In the case of a macroscopically isotropic aggregate it suffices to calculate the linear invariants C_{iijj} and C_{ijij} alone. In this case the averaging is performed automatically if one inserts for c_{ijkl} the elastic moduli of the main orientation (Kröner 1958).

For an aggregate of cubic crystals we have gone through the calculation. In this case the determination of the macroscopic compression modulus is trivial (it follows from C_{iijj}). Hence we only need to solve (5.18.28) for C_{ijij} (Kröner 1958). The result can be written in the form

$$G = \bar{\mu} - \frac{a x^2}{1 - x - 6x^2}$$

where

$$\bar{\mu} = \frac{3\mu + 2\nu}{5}, \quad a = 5\bar{\mu}\,\frac{3\bar{k} + 4\bar{\mu}}{\bar{k} + 2\bar{\mu}}, \quad x = \frac{6(\mu - \nu)}{5a}$$

and

$$\bar{k} = (c_{11} + 2c_{12})/3, \quad \mu = c_{44}, \quad \nu = (c_{11} - c_{12})/2$$

are the essential single crystal moduli. The above value of G is correct to the 3-rd order in x in the case of the perfectly disordered statistically isotropic and homogeneous aggregate of cubic crystallites.

d) Comments

Beside the physical assumptions which specify the physical range in which the concept of effective elastic moduli is valid we have used a few mathematical assumptions which we summarize now. In order to carry out the calculations we have required

(i) the differentiability of c_{ijkl} ,

(ii) the convergence of the iteration procedure,

(iii) the commutability of the averaging with the differentiations and integrations in eq. (5.18.13).

It can be seen from the mathematical theory of probability that (iii), actually, follows from (i). (*) Since in the final formulas e.g. in eq. (5.18.13), differentiations of c_{ijkl} do not occur but only differentiations of the macroscopic correlation functions the differentiation of which, in fact, have a meaning, we can make the transition from a continuous distribution of c_{ijkl} to one which is discontinuous without encountering any difficulty. From this we conclude, that our formulas apply also to the piecewise continuous distributions of elastic moduli in our problem. Then we only need to regard the convergence problem (ii). It is clear that the convergence of the unaveraged series is on-

---

(*) The admissibility of (iii) is, in fact, one of the most important consequences of the ergodic theorem.

ly ensured if the tensor c'_{ijkl} (for instance its eigenvalues) is sufficiently small, which means that the anisotropy of the crystallites is not too large.

It is possible that the averaged series (5.18.20) converges even in those cases where the unaveraged series does not converge. This could be concluded if we were able to prove that the truncated series (5.18.20) provides bounds which approach $\underline{C}$ if the number of included terms is increased. This problem to which we come back in section 5.20 has, indeed, a positive answer so that we can be sure of the convergence of the solution (5.18.20). Interestingly, it is the occurrence of the prime operation which makes the series convergent.

The result (5.18.28) is closely related to the self-consistent solution which was briefly described in section 5.16. In fact, this solution is obtained if the tensor $\underline{c}$ which appears on the right of eq. (5.18.20) through the tensor $\underline{c}'$ and $\underline{E}$ is replaced by $\underline{C}$. If we write both the self-consistent and the present solution as a power series in the relative fluctuation x of the shear modulus of an aggregate of cubic crystals, then the difference in the two results is of fourth degree in x. This fact explains why the self-consistent shear modulus is very close to the true effective modulus as long as the anisotropy is not too large (note that x is essentially the anisotropic constant of cubic crystals).

L. Thomsen has shown that the arithmetic Hill

average is nearly correct in the second power of x. This fact explains that also this average is often a good approximation.

Our calculations show that the n-th point correlation function of the aggregate of cubic crystallites is proportional to x^n. From this result we conlcude that the Hashin-Strikman bounds which are calculated with the help of the 2nd-order correlation functions (*) are correct up to x^2 and deviate from the true modulus by a term proportional to x^3. They contain within themselves the self-consistent modulus but not always the Hill average. On the other hand, the Voigt and Reuss bounds contain only the information about the 1st-order correlation and thus deviate by a term proportional to x^2. Considerations of this kind offer a rational way of estimating the reliability of the many approximations which have been proposed in the literature in the last 40 years.

In order to check the present and other theories one needs good experimental values of single crystals and polycrystals. Whereas good values are easily available for single crystals, it is quite difficult to interpret the result on polycrystals. In most cases one knows of which materials the crystallites consist, and perhaps, that the specimen is macroscopically isotropic. Already the latter statement is problematical because small deviations from isotropy alter the experimental results.

(*) Hashin and Shtrikman do not speak of correlation functions. They use information, however, which is essentially that of 2nd-order correlation functions.

But even if the specimens were isotropic with sufficient accuracy, we know from Hill's theory that the effective moduli can lie anywhere between the Voigt and Reuss bounds. Only further experimental information which could be extracted from X-ray measurements, for instance, can tell us the structural condition to which the measured moduli belong. Often the X-ray measurements are replaced by some vague statement that the specimen formed a "random" aggregate of crystallites. Deviations of this randomness from the perfect disorder of our theory again will cause deviations of the experimental moduli.

Finally, other effects have an influence on the measured values of elastic moduli. We mention especially the effect of the bowing out of dislocations which can falsify measurements of elastic moduli by a few percent. G. Bradfield and H. Pursey have discussed the way this error can be avoided by refined experimental techniques.

On account of all these sources of error, the measured elastic moduli of the so-called random polycrystalline aggregates can deviate from the true moduli of perfectly disordered aggregates by 10 % and more in unfavourable situations. Since the mutual differences between the calculated effective moduli, the self-consistent moduli and the Hill averages are mostly much smaller than 10%, it is very difficult to verify the validity of a particular theory from experimental results only.

Inspite of the above mentioned difficulty the literature is full of comparisons. We refer to the more recent work of O.L. Anderson (1965), W. Wawra (1970) and L. Thomsen (1971) from which one can conclude that the calculated elastic moduli are at least within the experimental accuracy.

We have not yet applied our results to the real composites, i.e. to multiphase and similar materials. It should not be a serious problem to do this. A further application could concern polycrystalline aggregates with texture which makes the polycrystal macroscopically anisotropic. In the idealized case of perfect fibre texture of an aggregate of cubic crystallites one has hexagonal symmetry, viz. perfect order in one direction and perfect disorder in the perpendicular sections. It seems that the solution of this problem can easily be obtained.

We conclude this section with a suggestion for the tabulating of elastic moduli of macroscopically isotropic polycristalline samples. We find it adequate to tabulate all exact bounds which correspond to the information given by the various perfect disorder correlation functions such as the bounds of Voigt and Reuss and those of Hashin and Shtrikman, but also those of the third order introduced in Section 5.20. As long as the exact effective moduli have not been obtained it seems sensible to tabulate the self-consistent values which lie even within the bounds of third order. Since the measurements of single crystal elastic moduli give very accurate results we obtain very accurate results for polycrystals with random distribution of the crystallites if the former results are used in our calculation.

5.19. The non-homogeneous case.

Theory of Beran and McCoy and of Mazilu (*)

a) Introduction

The effective elastic moduli calculated in the last section for the case of perfect order should be good enough to describe the response of the macroscopically homogeneous polycrystalline aggregate also in non-homogeneous stress situations provided, the volume forces and internal stresses are absent and the variations of stress and strain are slow enough to ensure that these quantities can be considered as macroscopically homogeneous in sufficiently large volume elements. Most of the applications of elasticity theory in static engineering problems are of this type.

The situation changes when the distribution of the local elastic moduli is no longer perfectly disordered. In this case correlations occur to which, in a rather rough way, we can assign a correlation length ℓ. The implication of this length is that the elastic moduli at two separate points $\mathbf{x}_1$ and $\mathbf{x}_2$, say, are correlated if the distance of these points is smaller

(*) *Note added in proof : W.M. Lewin (1971) as well as P. Dederichs and R. Zeller (1972) have independently arrived at very similar conclusions.*

than ℓ. The case of a finite correlation length ℓ has been treated by Beran and McCoy and by Mazilu. They have derived equations for the mean field variations and a mean stress -mean strain relation which is not of the form of an effective Hooke's law. In this description we adopt essentially the notation of Beran and McCoy which is quite different from that of Mazilu inspite of the closely related content of these works.

b) General theory

As in the last section we start from the deterministic equation

$$\partial_j (c_{ijk\ell}\, \varepsilon_{k\ell}) + F_i = 0 \tag{5.19.1}$$

where $c_{ijk\ell}(\underline{x})$ is a statistically homogeneous distribution. Following Beran and McCoy we take F_i to be dependent on position in a non-random way. For simplicity we assume that the body is infinite.

Decomposing the quantities of eq. (5.19.1) into ensemble means (*) and deviations we have

$$\bar{c}_{ijk\ell}\, \partial_j \bar{\varepsilon}_{k\ell} + \bar{c}_{ijk\ell}\, \partial_j \varepsilon'_{k\ell} + \partial_j (c'_{ijk\ell}\, \bar{\varepsilon}_{k\ell}) + $$

$$+ \partial_j (c'_{ijk\ell}\, \varepsilon'_{k\ell}) + F_i = 0 . \tag{5.19.2}$$

It differs from eq. (5.18.4) in the term F_i and the first member of eq. (5.19.2) which does not appear in the other equation. We

(*) *We emphasize that we use ensemble means rather than volume averages because we have a non-homogeneous situation.*

now allow the mean $\bar{\varepsilon}_{kl}$ to vary in space so that its derivative need not vanish. Of course, $\bar{\varepsilon}_{kl} = \bar{\varepsilon}_{kl}(\underline{x}) = \overline{\varepsilon_{kl}(\underline{x})}$ is a macroscopic function of position. We recall that $\bar{c}_{ijkl}$ does not depend on $\underline{x}$.

Starting from eq. (5.19.2) we can establish equations for the correlation functions of ε_{kl} (or ε'_{kl}) and the mixed functions of ε_{kl} and c_{ijkl}. The simplest equation is obtained by taking the ensemble average of (5.19.2). We observe the rule for commutation of average and differentiation. Thus we have

$$\bar{c}_{ijkl}\, \partial_j \bar{\varepsilon}_{kl} + \partial_j \overline{c'_{ijkl}\, \varepsilon'_{kl}} + F_i = 0 . \qquad (5.19.3)$$

If we subtract this from eq. (5.19.2) we derive the equation which governs the deviation part ε'_{kl} of the strain:

$$\bar{c}_{ijkl}\, \partial_j \varepsilon'_{kl} + \partial_j \left(c'_{ijkl}\varepsilon'_{kl}\right)' = -\partial_j \left(c'_{ijkl}\, \bar{\varepsilon}_{kl}\right) \qquad (5.19.4)$$

where $\left(c'_{ijkl}\, \varepsilon'_{kl}\right)' \equiv c'_{ijkl}\, \varepsilon'_{kl} - \overline{c'_{ijkl}\, \varepsilon'_{kl}}$. Eq. (5.19.4) is equal to eq. (5.18.4) if we assume a statistically homogeneous situation, in which, the derivatives of average vanisch.

Eq. (5.19.4) is now treated by Beran and McCoy in a very similar manner as we have treated the eq. (5.18.4), namely by an iterative sol‟tion in terms of the Green's tensor defined from $\bar{c}_{ijkl}$. The main difference is that Beran and McCoy do not use the spaces $\underline{x}_{12}$, $\underline{x}_{23}$ etc. because they allow for boundary conditions in which case they need a Green's tensor which depends on $\underline{x}_1$ and $\underline{x}_2$ separately, and not only on $\underline{x}_{12}$. This difference disappears if the body is chosen to be infinite.

There is no point in writing down the explicit expression for $\overline{c'_{ijkl}\varepsilon'_{kl}}$. For the further considerations it suffices to give the general form of the correct solution:

$$\overline{c'_{ijkl}\,\varepsilon'_{kl}} = A_{ijkl}\,(\underline{x}_1,\underline{x}_2)\,\bar{\varepsilon}_{kl}(\underline{x}_2) \tag{5.19.5}$$

where $A_{ijkl}(\underline{x}_1,\underline{x}_2)$ represents an infinite sum of integro-differential operators. Beside the Green's tensor this operator contains the infinite set of correlation functions of c'_{ijkl}.

If we substitute eq. (5.19.5) in eq. (5.19.3) we obtain

$$\bar{c}_{ijkl}\,\partial_j^{(1)}\bar{\varepsilon}_{kl}(\underline{x}_1) + \partial_j^{(1)}\left[A_{ijkl}\,(\underline{x}_1,\underline{x}_2)\,\bar{\varepsilon}_{kl}(\underline{x}_2)\right] + F_i(\underline{x}_1) = 0 \tag{5.19.6}$$

which is essentially Beran and McCoy's equation for $\bar{\varepsilon}_{kl}$.

Let us compare the last result with the equation which follows from the local equilibrium equation

$$\partial_j^{(1)}\,\sigma_{ij}\,(\underline{x}_1) + F_i\,(\underline{x}_1) = 0. \tag{5.19.7}$$

Since F_i is non-random we have

$$\partial_j^{(1)}\,\bar{\sigma}_{ij}(\underline{x}_1) + F_i\,(\underline{x}_1) = 0 \tag{5.19.8}$$

and a comparison with eq. (5.19.6) shows that

$$\bar{\sigma}_{ij}(\underline{x}_1) = \bar{c}_{ijkl}\,\bar{\varepsilon}_{kl}\,(\underline{x}_1) + A_{ijkl}\,(\underline{x}_1,\underline{x}_2)\,\bar{\varepsilon}_{kl}\,(\underline{x}_2). \tag{5.19.9}$$

The mean stress - mean strain relation is now a functional equation between $\bar{\underline{\sigma}}$ and $\bar{\underline{\varepsilon}}$. If the body is infinite then the differential part of the operator A_{ijkl} can be removed by partial integration giving A_{ijkl} and eq. (5.19.9) can be reduced to (*)

$$\bar{\sigma}_{ij}(\underline{x}_1) = C_{ijkl}\,\bar{\varepsilon}_{kl}(\underline{x}_1) + \\ + \int a_{ijkl}(\underline{x}_1-\underline{x}_2)\,\bar{\varepsilon}_{kl}(\underline{x}_2)\,dV_2 \qquad (5.19.10)$$

where we have combined the local part of $A_{ijkl}(\underline{x}_1=\underline{x}_2)$ with $\bar{c}_{ijkl}$ which results in the formerly calculated effective moduli C_{ijkl}. The non-local part depends on $\underline{x}_1 - \underline{x}_2$ alone because the material properties were assumed to be macroscopically homogeneous.

Eqs. (5.19.9, 5.19.10) show that in a body with macroscopically homogeneous distribution of elastic moduli the mean stress is completely determined by the mean strain, given the macroscopic material description in terms of C_{ijkl} and a_{ijkl} (or $\bar{c}_{ijkl}$ and A_{ijkl}) or in terms of the set of correlation functions. The macroscopic stress-strain law is a typical non-local relation of the form encountered in non-local elasticity theory. In fact, we have obtained a realization of a non-locally elastic material. This realization can well be of practical importance. For non-local elasticity theory including its

(*) Beran and McCoy derive this equation for the infinite, locally isotropic body. Our consideration shows that eq. (5.19.9) and, in particular, eq. (5.19.10) apply more generally.

relation to strain gradient theories we refer the reader to the references of this chapter.

Physically it is clear that the "range" of the non-local kernel $a_{ijkl}(\underline{x}_1-\underline{x}_2)$ is of the order of the correlation length l. If we make this smaller and smaller we gradually approach the local situation. Obviously, this is the case of perfect disorder treated in section 5.18.

The most important result of this investigation is the discovery that the linear non-local elasticity theory can be mapped onto the linear elasticity theory of heterogeneous materials formulated for the averages of stress and strain.

This opinion is also expressed by Dr. Mazilu in a communication to the present author on 21st of November 1969: "J'ai le plaisir de vous annoncer que, conformement aux résultats obtenus dans ce travail, les milieux "parfaitement desordonnés" considerés pour la première fois par vous, constituent une classe remarcable de milieux statistiquement homogènes; ils sont les uniques milieux pour lesquels les "lois constitutives" macroscopiques sont locales et ne dépendent de la form géometrique du corps."

This citation shows clearly the important contribution of Mazilu.

The influence of the boundary has been discussed by Beran and McCoy in some detail. It seems rather obvious that in the case of perfectly disordered materials the boundary does not offer particular difficulties. The reason is, of course, that in this case the correlation length ℓ is zero: there is no boundary layer effect. This is clearly expressed by Mazilu's statement.

5.20. Recent results

a) Relation to quantum mechanical scattering theory

In the interval between the writing this manuscript and the proof-reading a number of new relations have been discovered of which we now give a very brief account.

P. Dederichs and R. Zeller (1972) start from Fredholm's formula (the analogon of Kirchhoff's formula in the potential theory) which, using the notations of the previous sections, reads

$$u_m(\underline{x}) = \int_V dV' G_{mi}(\underline{x},\underline{x}') F_i(\underline{x}') - \int_S dS'_j c^o_{ijkl} G_{mk,l'}(\underline{x},\underline{x}') u_i(\underline{x}') + \int_S dS' G_{mi}(\underline{x},\underline{x}') A_i(\underline{x}') \tag{5.20.1}$$

where G_{mi} is the Green's tensor derived form the arbitrary constant tensor c^o_{ijkl}. They note that in this formula G_{mi} can be either the (self-adjoint) Green's tensor $\underline{G}^{(\infty)}$ of the infinite medium or one of the (self-adjoint) tensors $G^{(1)}$, $G^{(2)}$, $G^{(m)}$ of the 1st, 2nd and mixed boundary value problem respectively, as can be seen from the usual derivation of this formula.

Splitting now our former $\underline{c}$ not, as we did, into $\bar{\underline{c}}$ and $\underline{c}'$, but more generally, according to

(5.20.2) $$\underline{c} = \underline{c}^0 + \delta \underline{c}$$

with $\underline{c}^0$ arbitrary constant, the equilibrium equations in absence of real volume forces are

(5.20.3) $$c^0_{ijkl}\, \partial_j\, \varepsilon_{kl} + \partial_j (\delta c_{ijkl}\, \varepsilon_{kl}) = 0$$

Treating the 2nd term as a fictive volume force density $F_i(\underline{x}')$ and using (5.20.1) appropriately we obtain after differentiation of u_m

(5.20.4) $$\varepsilon_{mn}(\underline{x}) = \int dV'\, \partial'_j \left(\delta c_{ijkl}(\underline{x}')\varepsilon_{kl}(\underline{x}')\right) G_{mi,n}(\underline{x},\underline{x}')\Big|_{(mn)} + \text{surface integrals}$$

where (mn) means symmetrization in m , n . After partial integration in ∂'_j and using now our previous abridget notation we find

(5.20.5) $$\underline{\varepsilon} = \underline{\varepsilon}^0 - \underline{\Gamma}\, \delta\, \underline{c}\, \underline{\varepsilon}$$

where Γ is the (self-adjoint) tensor function defined as in (5.18.15), and

(5.20.6) $$\varepsilon^0_{mn}(\underline{x}) = \int_S u_i(\underline{x}')\, c^0_{ijkl}\, \Gamma_{klmn}(\underline{x},\underline{x}')\, dS'_j + \int_S A_i(\underline{x}')\, G_{mi,n}(\underline{x},\underline{x}')\, dS'$$

with $A_i = n_j c_{ijkl}\,\varepsilon_{kl}$. Also here G_{mi} can be any of the Green's tensors mentioned above. Of course, $\underline{G}$ and $\underline{\Gamma}$ are to be formed with $\underline{c}^{\circ}$ rather than with $\bar{\underline{c}}$.

Eq. (5.20.5) which is an exact equation of the elasticity theory of heterogeneous materials has the form of the celebrated Lippmann-Schwinger equation of the quantum mechanical scattering theory. This is a very pleasant feature because the well-developed methods of this theory can now be taken over to the present problems as described in detail in the work of Dederichs and Zeller (1972).

We now apply eq. (5.20.5) to our statistical problem by assuming that Hill's condition is valid. Then we can prescribe the displacements or the forces (or partly forces-partly displacements) on the surface as smooth which apparently results in a smooth $\underline{\varepsilon}^{\circ}$ (see eq. 5.20.6). Taking $\underline{c}^{\circ} = \bar{\underline{c}}$, $\delta\underline{c} = \underline{c}'$ we obtain

$$\underline{\varepsilon}' = -\underline{\Gamma}\,P\,\underline{c}'\underline{\varepsilon} \equiv -\underline{\Gamma}\,\underline{c}'\bar{\underline{\varepsilon}} - \underline{\Gamma}\,P\,\underline{c}'\underline{\varepsilon}' \tag{5.20.7}$$

which is our fundamental equation (5.18.14). After this the procedure is as usual. In particular, eq. (5.18.20) remains the solution of the problem.

The last consideration shows that in Γ we can use, as we wish, the Green's tensor of the 1st, 2nd or mixed boundary problem. Dederichs and Zeller give an estimation according to which also the tensor of the infinite medium with an elastic tensor $\bar{\underline{c}}$ can be used if the grain number tends to infinity. (Question to the reader: Why does this statement need an additional consideration?)

b) Effective compliances

If the Lippmann-Schwinger equation for the strain (5.20.5) is multiplied by $\underline{c}^0$ and the quantities $\underline{\sigma}^0 = \underline{c}^0 \underline{\varepsilon}^0$, $\underline{s}^0 = (\underline{c}^0)^{-1}$, $\delta\underline{s} = \underline{s} - \underline{s}^0$ are introduced one easily obtains the Lippmann-Schwinger equation for the stress

(5.20.8)
$$\underline{\sigma} = \underline{\sigma}^0 - \underline{\Delta}\,\delta\underline{s}\,\underline{\sigma}$$

where

(5.20.9)
$$\underline{\Delta} = \underline{c}^0 - \underline{c}^0 \underline{\Gamma}\,\underline{c}^0$$

We put $\underline{s}^0 = \bar{\underline{s}}$, then $\delta\underline{s} = \underline{s}'$, and after performing the prime operation on eq. (5.20.8) we obtain

(5.20.10)
$$\underline{\sigma}' = -\underline{\Delta}\,P\underline{s}'\underline{\sigma} = -\underline{\Delta}\,\underline{s}'\underline{\sigma} - \underline{\Delta}\,P\underline{s}'\underline{\sigma}'$$

where

(5.20.11)
$$\underline{\Delta} = \bar{\underline{s}}^{-1} - \bar{\underline{s}}^{-1}\underline{\Gamma}\,\bar{\underline{s}}^{-1}$$

and $\underline{\Gamma}$ is calculated with $\bar{\underline{s}}^{-1}$ rather than with $\bar{\underline{c}}$. We now can write down the two rigorous equations

(5.20.12)
$$\underline{C} = \bar{\underline{c}} - \overline{\underline{c}'\underline{\Gamma}P\underline{c}'} + \overline{\underline{c}'\underline{\Gamma}P\underline{c}'\underline{\Gamma}P\underline{c}'} - \overline{\underline{c}'\underline{\Gamma}P\underline{c}'\underline{\Gamma}P\underline{c}'\underline{\Gamma}P\underline{c}'} + \dots$$

$$\underline{S} = \overline{\underline{s}} - \overline{\underline{s}'\underline{\Delta}P\underline{s}'} + \overline{\underline{s}'\underline{\Delta}P\underline{s}'\underline{\Delta}P\underline{s}'} - \overline{\underline{s}'\underline{\Delta}P\underline{s}'\underline{\Delta}P\underline{s}'\underline{\Delta}P\underline{s}'} + \dots \quad (5.20.13)$$

These equations are equivalent to those derived by Dederichs and Zeller (1972) who, however, did not use the prime operator. W.M. Lewin (1971) and P. Mazilu (1972) derived equations which seem to be equivalent to (5.20.12). Incidentally, the series (5.20.13) with the P missing by mistake - hence only correct to the 3rd order - had been derived by S.L. Koh, H. Koch and myself in 1969 by use of the method of 3-dimensional stress functions (unpublished work).

The first three terms in the above equations are easily calculated in the case of macroscopically isotropic aggregates of cubic crystallites. In this approximation one obtains

$$2G = 2\left[\bar{\mu} - 3E\varphi^2 - 3E^2\varphi^3 + O(\varphi^4)\right] \quad (5.20.14)$$

$$(2G)^{-1} = 2\left[\bar{M} - 3F\Phi^2 - 3F^2\Phi^3 + O(\Phi^4)\right] \quad (5.20.15)$$

where

$$E = \frac{3}{5\bar{\mu}} \cdot \frac{k + 2\bar{\mu}}{3k + 4\bar{\mu}}, \quad \bar{\mu} = \frac{3\mu + 2\nu}{5}, \quad \varphi = \frac{2(\mu - \nu)}{5}$$

$$F = \frac{3}{5\bar{M}} \cdot \frac{K' + 2\bar{M}}{3K' + 4\bar{M}} , \quad \bar{M} = \frac{3M + 2N}{5} , \quad \Phi = \frac{2(M-N)}{5}$$

$$K'k = 4/9 , \quad M\mu = N\nu = 1/4$$

and

$$k = \frac{c_{11} + 2c_{12}}{3} , \quad \mu = c_{44} , \quad \nu = \frac{c_{11} - c_{12}}{2}$$

are the essential elastic moduli of the cubic crystal (*). F follows from $\underline{\Delta}$ in the same way as $E\ (=2E_{1212})$ followed from $\underline{\Gamma}$ (cf. section 5.18). In the next paragraph we show that the G's calculated from eqs. (5.20.14) and (5.4.15) by omitting the parts $O(\varphi^4)$ and $O(\Phi^4)$ are upper and lower bounds to the true effective shear modulus.

c) Bounds for the effective elastic moduli

The following method for finding bounds is, again, due to Dederichs and Zeller (1972). We give here a somewhat abridged derivation. Because the local tensor $\underline{c}$ and $\underline{s}$ of elastic moduli and compliances, respectively, are positive definite, we have

---

(*) If the terms $O(\varphi^3)$ are omitted in eq. (5.20.14) one obtains a result alreadv derived by Lifshitz and Rosensveig in 1946.

$$\int_V (\underline{\varepsilon} - \underline{\varepsilon}^*)\, \underline{c}\, (\underline{\varepsilon} - \underline{\varepsilon}^*)\, dV \geq 0 \,, \quad \int_V (\underline{\sigma} - \underline{\sigma}^*)\, \underline{s}\, (\underline{\sigma} - \underline{\sigma}^*)\, dV \geq 0$$

(5.20.16)

where $\underline{\varepsilon}^*$ and $\underline{\sigma}^*$ are arbitrary strain and stress tensors respectively. Assume that

$$\int_V \underline{\varepsilon}\,\underline{\sigma}\, dV = \int_V \underline{\varepsilon}^*\underline{\sigma}\, dV \,, \quad \int_V \underline{\sigma}\,\underline{\varepsilon}\, dV = \int_V \underline{\sigma}^*\underline{\varepsilon}\, dV$$

(5.20.17)

where $\underline{\sigma} = \underline{c}\,\underline{\varepsilon}$ and $\underline{\varepsilon} = \underline{s}\,\underline{\sigma}$. Then

$$\int_V \underline{\varepsilon}^*\underline{c}\,\underline{\varepsilon}^*\, dV \geq \int_V \underline{\varepsilon}\,\underline{c}\,\underline{\varepsilon}\, dV \,, \quad \int_V \underline{\sigma}^*\underline{s}\,\underline{\sigma}^*\, dV \geq \int_V \underline{\sigma}\,\underline{s}\,\underline{\sigma}\, dV \,. \qquad (5.20.18)$$

These are the two variational principles from which the bounds for the effective elastic moduli are derived. In order to apply these principles we have to establish the conditions under which the assumptions (5.20.17) are valid.

Regarding the symmetry of $\underline{\sigma}$ we replace in eqs. (5.20.17) $\underline{\varepsilon}$ by $\underline{\nabla}\,\underline{u}$ and $\underline{\varepsilon}^*$ by $\underline{\nabla}\,\underline{u}^*$ and integrate partially:

$$\int_S dS\, \underline{A}\cdot\underline{u} - \int_V dV\, \underline{F}\cdot\underline{u} = \int_S dS\, \underline{A}\cdot\underline{u}^* - \int_V dV\, \underline{F}\cdot\underline{u}^* \qquad (5.20.19)$$

$$\int_S dS\, A\cdot u - \int_V dV\, \underline{F}\cdot\underline{u} = \int_S dS\, A^*\cdot u - \int_V dV\, \underline{F}^*\cdot\underline{u}\,. \qquad (5.20.20)$$

Hence the conditions for the 1st principle are

$$\left\{\underline{\varepsilon} = \mathrm{Def}\,\underline{u}\,,\ \underline{\varepsilon}^* = \mathrm{Def}\,\underline{u}^*,\ \underline{F} = 0\right\} \text{ in volume, } \left\{\underline{u} = \underline{u}^*\right\} \text{ on the boundary}$$

(5.20.21)

and those for the second principle are

$$\{\underline{\varepsilon} = \mathrm{Def}\,\underline{u},\ \underline{F} = \underline{F}^* = 0\}\ \text{in volume},\ \{\underline{A} = \underline{A}^*\}\ \text{on the boundary}$$
(5.20.22)

Here $(\mathrm{Def}\,\underline{u})_{ij} \equiv (\partial_i u_j + \partial_j u_i)/2$. Hence the manifold of the functions $\underline{\varepsilon}^*$ and $\underline{\sigma}^*$ admitted for comparison in the variation are those which satisfy (5.20.21), but not $\underline{F} = 0$ and those obeying (5.20.22) but not the compatibility equations respectively.

We now write

$$(5.20.23)\quad \underline{\varepsilon}^* = \Big[\underline{I} + \sum_{m=1}^{N} (-\underline{\Gamma} P \underline{c}')^m\Big]\bar{\underline{\varepsilon}} = \bar{\underline{\varepsilon}}\Big[\underline{I} + \sum_{m=1}^{N} (-\underline{c}'\underline{\Gamma} P)^m\Big]$$

$$(5.20.24)\quad \underline{\sigma}^* = \Big[\underline{I} + \sum_{m=1}^{N} (-\underline{\Delta} P \underline{s}')^m\Big]\bar{\underline{\sigma}} = \bar{\underline{\sigma}}\Big[\underline{I} + \sum_{m=1}^{N} (-s'\underline{\Delta} P)^m\Big]$$

where $N = 0,1,2,\ldots$. If $\underline{\Gamma}$ is formed with the Green's tensor $\underline{G}^{(1)}$ then $\underline{\varepsilon}^*$ satisfies the conditions (5.20.21). Conversely, if $\underline{\Delta}$ is formed with $\underline{G}^{(2)}$ then $\underline{\sigma}^*$ satisfies the conditions (5.20.22). Introducing the effective tensors $\underline{C}$ and $\underline{S}$ in eqs. (5.20.18) gives us

$$(5.20.25)\quad \int \underline{\varepsilon}^* \underline{c}\, \underline{\varepsilon}^* dV \geq \int \bar{\underline{\varepsilon}}\, \underline{C}\, \bar{\underline{\varepsilon}}\, dV,\quad \int \underline{\sigma}^* \underline{s}\, \underline{\sigma}^* dV \geq \int \bar{\underline{\sigma}}\, \underline{S}\, \bar{\underline{\sigma}}\, dV.$$

Inserting eqs. (5.20.23) and (5.20.24) into the averaged equations (5.20.25) respectively we obtain

$$(5.20.26)\quad \overline{\Big[\underline{I} + \sum_{m=1}^{N} (-\underline{c}'\underline{\Gamma} P)^m\Big]\underline{c}\Big[\underline{I} + \sum_{m=1}^{N} (-\underline{\Gamma} P \underline{c}')^m\Big]} \geq \underline{C}$$

$$\overline{\left[\underline{I}+\sum_{m=1}^{N}\left(-\underline{s}'\underline{\Delta}P\right)^{m}\right]\underline{s}\left[\underline{I}+\sum_{m=1}^{N}\left(-\underline{\Delta}P\underline{s}'\right)^{m}\right]} \geq \underline{S}\ . \qquad (5.20.27)$$

In order to calculate the left sides in eqs. (5.20.26) and (5.20.27) one uses with advantage the relations

$$\underline{\Gamma}\,\underline{\bar{c}}\,\underline{\Gamma} = \underline{\Gamma}\ ,\quad \underline{\Delta}\,\underline{\bar{c}}\,\underline{\Delta} = \underline{\Delta} \qquad (5.20.28)$$

where the second one follows easily from the first one. To prove the first (5.20.28) one observes that according to eq. (5.18.16) the tensor function $\underline{\Gamma}$ formed from $\underline{\bar{c}}$ can be understood as the operator which transforms any stress field $\underline{\sigma} = \underline{\bar{c}}\,\underline{\varepsilon}$ into the strain field $\underline{\varepsilon}$. Hence

$$\underline{\Gamma}\,\underline{\bar{c}}\,\underline{\Gamma}\,\underline{\sigma} = \underline{\Gamma}\,\underline{\bar{c}}\,\underline{\varepsilon} = \underline{\Gamma}\,\underline{\sigma}$$

This proves eqs. (5.20.28).

The above conclusions apply if $\underline{\Gamma}$ is formed with the Green's tensor $\underline{G}^{(2)}$. Dederichs and Zeller give a proof which is valid for $\underline{G}^{(1)}$. Because the Green's tensors $\underline{G}^{(\infty)}$, $G^{(1)}$, $G^{(2)}$, $G^{(m)}$, are allowed in the solutions (5.20.12) and (5.20.13) it should suffice to have one proof.

If now we write $\underline{c} = \underline{\bar{c}} + \underline{c}'$, $\underline{s} = \underline{\bar{s}} + \underline{s}'$ in eqs. (5.20.26) and (5.20.27) respectively and calculate the left sides, using eqs. (5.20.28), for $N = 0,1,2...$, we find expressions which are identical with the solutions (5.20.12) and (5.20.13) if we truncate these series after $n = 1,3,5...$ terms.

It follows that this procedure of cutting short the series in these solutions yields upper bounds to the effective elastic moduli and compliances respectively. Since $\underline{S} = \underline{C}^{-1}$ we find at the same time lower bounds for the effective elastic moduli (or compliances, if we wish). These bounds are of order $n = 1, 3, 5, \ldots$, i.e. correct in the 1st, 3rd, 5th ... degree of the fluctuations $\underline{c}'$ (or $\underline{s}'$). No such bounds have yet been established for $n = 2, 4, 6, \ldots$ We recall that the bounds of Hashin and Shtrikman which are correct in the 2nd degree of the fluctuations follow from different variational principles.

In order to see how the bounds close up we show in fig. 1 their values for $n = 1$ (Voigt - Reuss), for Hashin and Shtrikman and for $n = 3$ in the case of the shear modulus G of polycristalline, macroscopically isotropic copper. In this case the grains are of cubic symmetry so that the formulas (5.20.14) and (5.20.15) can be applied. It appears that the progress made since the old work of Voigt and Reuss is remarkable.

Figure 1

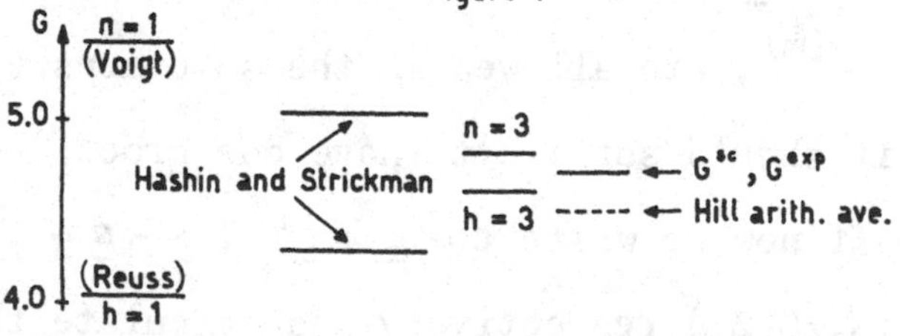

Caption of Fig.1.

Bounds and other results for the effective elastic moduli of perfectly disordered, macroscopically isotropic c11 **polycrystalline copper. The values for the bounds of n = 3 are 4.76 and 4.87 respectively. The experimental and self-consistent values coincide within experimental accuracy at 4.83 (all values is 10^{11} dyn/cm^2. All experimental results used were taken from G. Bradfield, quoted by G. Kneer, l.c.**

That the self-consistent value G^{sc} lies within the narrow bounds of $n = 3$ is not accidental. It has been proved by the application of a method developed by E. Kröner, B.K. Datta and D. Kessel (1966) that G^{sc} always lies between the bounds of 3rd order, a result which shows that the self- consistent values are particularly useful.

d) Deviations from Hill's condition

This subject is treated by E. Kröner (1973). The situation is much more complex. I restrict myself to a reproduction of the abstract of my paper in 1973:

"The complete macroscopic description of the behaviour of bodies with a microscopically nonuniform, at least partially disordered constitution requires an infinite number of functions. This purpose is served by the correlation functions of the material tensor as well as those of the relevant field and source quantities. The crucial importance of the degree of order of the constitution is demonstrated through the example of linear elastostatics. It is assumed that the body behaves locally according to the laws of convantional elasticity theory.

The behaviour of such "random media" can be classified into four categories which we refer to as primitive local, primitive non-local, non-primitive local and non-primitive non-local. The term primitive implies that the mean fields can be calculated individually from a closed system of equations without employing correlation functions. For this to be applicable it is necessary and sufficient that Hill's condition $(\sigma\varepsilon)=(\sigma)(\varepsilon)$ be valid, where σ, ε, are respectively stress and strain tensor. This situation arises in polycristal and multi-phase materials if, and only if, (1) an ergodic hypothesis can be established, (2) the number of grains is infinite, and (3) distributions of volume stress sources, if present, are not correlated to the local elastic moduli.

If the material parameters are correlated over finite distances one obtains a non-local response law which is of the standard form in the primitive situation. Only if the constitution of the body is perfectly disordered and the situation primitive do the mean fields obey a response law with an effective material tensor.

If all correlation functions of the local material parameters are known then a set of average Green's functions can be calculated which allows to express the desired macroscopic fields in terms of the macroscopic sources."

In addition, the paper contains a formal solution of the problem in the language of functionals. This solution gives the full macroscopic information in form of the combined characteristic functional M of $\underline{\varepsilon}$, $\underline{c}$, $\underline{F}$, $\underline{A}$ which is expressed by the assumed-as-given probability density functional P of $\underline{c}$, $\underline{F}$, $\underline{A}$. For details the reader is referred to the original work.

CHAPTER VI

THE STATISTICAL PROBLEM OF PLASTICITY

6.21. Introduction. Plasticity and viscoplasticity

From a phenomenological standpoint the difference between the two properties of solids which enable permanent deformation, namely plasticity and viscosity, consists mainly in the following : Whereas plastic materials show only a weak temperature dependence one observes strong temperature effects in viscous deformation. To this thermal behaviour of the two kinds of materials corresponds an inherent time behaviour: plastic deformation is time-independent in the sense of inherent times whereas viscous behaviour is strongly time-dependent.

In addition to the above effects one observes that plastic deformation causes remarkable changes in the resistance against further deformation - this is the phenomenon of work-hardening - whereas viscous materials, when deformed, do not show indications of a change of their internal conditions.

All the above statements are somewhat idealized. In fact, there are no purely plastic (or elastoplastic) or purely viscous (or visco-elastic) solids. Plasticity is always accompanied by some amount of viscosity and vice versa. For this reas on our statements are true only in an idealized sense; notwithstanding, we can define plasticity and viscosity so that the statements are true. On the other hand we can classify plasticity and viscosity from a microscopic standpoint. Using again a little bit of simplification we state that plastic deformation is the result of the motion of line-shaped objects whereas viscous deformation results from moving point-objects. Although this difference seems to be mainly one of geometry it has profound consequences on the physical behaviour of such materials. One of the main reasons is that thermal activation can help the motion of such objects only if these are concentrated in a volume comprising of not more than a few atoms or molecules, and this case is realized only with the point-shaped objects. This remark immediately explains the difference in time and temperature behaviour of plastic and viscous materials.

More important in our present study is the internal state behaviour. The observed change of the internal state in the case of plastic deformation can be attributed to a multiplication of the line objects. This multiplication at the same time results in new arrangements of the lines. Since plastic deformation is the motion of these lines, the change of their arrangement certainly causes a change of the resistance against further plastic deformation, hence a change in the internal state of the body. A corresponding effect is not observed during viscous deformation because the point objects do not multiply themselves.

Having so classified, at least to some degree, the difference between plastic and viscous behaviour we now restrict ourselves to plastic materials. The line objects are known as dislocations in solid state physics or materials science. They are disturbances (defects) of the otherwise regular crystalline structure of plastic materials. (Viscous behaviour is only encountered if materials contain certain amorphous rather than crystalline elements.)

The theory of dislocations is an extended branch of solid state physics to which the reader is referred. Here we can only mention the second important feature of the dislocations beside being the source of plastic deformation: Dislocations produce around themselves a long range stress and strain field through which they interact mutually. This interaction determines the plastic behaviour of the material in a fundamental manner.

We give the reader an idea of the numbers involved: A typical number of dislocations piercing through 1 cm^2 of an undeformed metallic material is 10^8. This number increases during plastic flow and reaches 10^{12} or even more after large deformation.

6.22. Preliminaries on statistical theories of plasticity

a) Turbulence-like theory

The stress and strain field around a dislocation

reaches very high values near the dislocation line and falls off inversely proportional to the first power of the distance in the special case of an infinite straight dislocation line. The numbers of dislocations quoted above imply that the mean distance between neighbouring dislocations varies from 10^{-4} cm in the undeformed material down to 10^{-6} cm and less in the considerably deformed material. Obviously, these lengths supply a scale for the fluctuations of the internal stress and strain in the body.

This way of speaking may be good enough for our more qualitative description. In quantitative investigations one has to observe that dislocations tend to form groups among themselves. Approximately one can take this into account by introducing two characteristic lengths, the first giving the mean distance between neighbouring dislocations within one group and the second informing about the mean distance between two neighbouring groups.

The result that a plastically deformed body is filled with internal stress and strain fields which fluctuate heavily on a microscopic scale very much reminds of the situation of turbulence where velocity fluctuations on a microscopic scale influence the behaviour of the fluid. This correspondence suggests us to formulate equations for stress and strain correlations just in the way we did for the turbulence problem. There is a decisive difference, however, between the two situations: In the case of turbulence we could base our statistics on a deterministic equation, the Navier-Stokes equation, which governs the material behaviour on the microscopic scale. In particular, this equation contains the material information in the form of the viscosity constant.

In the case of plasticity we do have the equation of motion in the form of eq. (3.12.1):

$$\varrho \frac{d\underline{v}}{dt} + \operatorname{div} \underline{\sigma} + \varrho \underline{F} = 0 . \qquad (6.22.1)$$

We do not possess a reliable description for the fact that it is the motion of a plastic material which is to be described. Hence we cannot replace the stress in (6.22.1) by a kinematical quantity via a material-characteristic law (constitutive law). The laws often used in the phenomenological theory of plasticity do not serve this purpose because they should rather come out from the statistical theory than being an a priori material information. This discussion shows clearly which preliminary investigation is necessary before the formalism of the statistical theory can be applied. Certainly, the needed material information should also contain something like a yield condition.

With these vague intimations about a field which, so far, lacks concrete results we conclude our discussion on a turbulence-like theory of plasticity.

b) Statistical theory of dislocations

We have emphasized in the introduction that the variation of the internal state of the body is characteristic of plastic deformation. The description of this internal state therefore has to be a basic problem of any theory of plasticity. In the turbulence-like approach discussed in the last section the internal state were to be described in terms of the stress and strain fluctuations. To use correlation functions in these quantities could well be the best way to do this.

Since the stress and strain fluctuations are in fact determined by the dislocation arrangements - we omit complications which imply that this statement is only approximate - we also can describe the internal state in terms of dislocations.

Many workers in the field of plasticity seem to agree that a satisfactory theory of plasticity can only be formulated if either internal state variables or quantities which can be mapped onto such variables are introduced. The critical point is always the physical interpretation of such internal state variables. As I have argued in 1963 all quantities describing dislocation arrangements can be used as internal state variable. They fulfill the requirement, which is by definition laid upon any state variable, that these are quantities which can be measured at a time without knowing anything about the past. I have, at the same time, argued that neither the so-called plastic work nor the second strain invariants are good state quantities in this sense and therefore should not be used as internal state variables in this theory.

Both the lack of information about the dislocation

arrangements and their stochastic nature suggests strongly a statistical description. Since we intend a description in macroscopic terms it is more suitable for our purpose to use the correlation functions of the dislocation distribution. A complete description requires the full set of correlation functions. Thus the number of internal state variables is infinite, a result which had been surmised by Drucker as early as 1960.

Little is known about the statistics of rigid lines; even less material exists on the statistics of deformable lines which, in addition, have to be allowed to intersect each other as real dislocations do. A particular feature of such a statistics has to be that this statistics is continuous in one dimension and discrete in the other two dimensions of the 3-dimensional space.

Our procedure of forming correlation functions remains valid, of course. In order to apply it we first need a mathematical description of dislocation lines. Such a description is provided by the continuum theory of dislocations. From this we know that the dislocation line element at some point $\underline{x}$ is fully characterized by two vectors, viz. the Burgers vector $\underline{b}$ and the vector $\underline{t}$ of the line element. This means that the dislocation line element has the mathematical character of a dyadic, i.e. a 2nd-rank tensor.

Taking the ensemble average over the dyadics $t_i b_j$ we obtain the average distribution of dislocations first introduced by J.F. Nye in 1953, and called the tensor α_{ij} of dislocation density. This dislocation density provides a rather poor information in most situations of interest. The reason is that dislocations usually do not form perfect line densities as, for example, current densities in electrodynamics often do. Mostly dislocations form loops with an average radius of curvature which is comparable to the mean distance of neighbouring dislocations.

The information supplied by the 2nd-order correlation function, which obviously is a 4th-rank tensor function, seems to be of much greater relevance. In my 1970 paper on this subject I have argued that one of the two linear invariants of this correlation function, taken for $\underline{x}_1 = \underline{x}_2$ has the meaning of the total line length of dislocations within a unit volume. This is a quantity of real physical interest. In fact, the existing physical theories which explain plastic behaviour on a microscopic and partly atomic scale make wide use of the total line length of dislocations. Obviously this is a true internal state variable.

The 2nd-order correlation tensor also accounts for dislocation dipoles which play an important role in some physical theories of plasticity.

These results show that the dislocation correlation functions are quantities which appear in microscopic theories, and could well be used in macroscopic theories, too. Thus they can help to bridge the often complained gap between the micro- and macro-approaches to plastic behaviour.

So far, the dislocation correlations have only descriptive character. No governing equations are known for them. One can, however, use them as internal state variables in a deterministic theory of plasticity. There they could replace inadmissible variables, such as the plastic work, and so lead to a theory closer to physical reality. Approaches in this direction have been made by J.Zarka(1969)1.c.and by C.Teodosiu(1970) (*) who use dislocation lengths in the various glide systems of a crystal as internal state variables of a deterministic theory. The authors have in addition established certain equations of evolution for these quantities on the basis of some elementary processes on the microscopic scale. It seems that this new direction is promising.

The general impression is, however, that the statistical theory of dislocation is still in a very preliminary stage.

c) Plastic deformation of heterogeneous materials

So far, the statistics has entered our investigation on the basis of random distributions of dislocations. If we now consider heterogeneous materials into which we also include polycrystalline aggregates, then new statistical elements come in. Assume, we had solved the plasticity problem of the

(*) Lecture in Oberwolfach, Germany, Febr. 1971. Cf. Also E. Kröner and C. Teodosin (1973)

single constituents. An additional statistical theory is then needed in order to combine the results on the constituents with a theory of the plastic behaviour of heterogeneous materials. This problem resembles the corresponding problem of elastic materials. There are complications, however, because grain boundaries are often impenetrable obstacles to dislocations. Not much is known about the macroscopic plasticity theory of heterogeneous materials, but it appears that statistics again would provide valuable tools to this end.

In this chapter we have only discussed very simple plastic materials. Many further complications arise if one considers solid solutions, order - disorder effects, recrystallization, and fatigue etc. . All these phenomena bear stochastic features and it is believed that probability theory and statistics can be of great assistance in problems of this kind. This field of material behaviour is, however, much too complex to speak about it in any detail.

APPENDIX

The following two tables put together by B.K. Datta and R. Bauer show for a number of macroscopically homogeneous and isotropic, perfectly disordered polycrystalline aggregates the Voigt bounds K_V, G_V and Reuss bounds K_R, G_R of the compression and shear moduli respectively, furthermore the asithmetic means K_{VRH}, G_{VRH} of these quantities and, finally, the effective compression and shear moduli K* G* calculated from eg. ( ) of section 18 by means of a computer programme. All single crystal values were taken from B.O. Anderson, l.c., with the exception of those of graphite for which we used values of E.J. Seldin and C.W. Nezbeda, l.c., The tables refer to the discussion of section 18, part d.

Table I

crystal class: cubic
all moduli in kilobars

Material	K*	G_V	G_R	G_{VRH}	G*	Temperature (°K)
Ag	1087	375	284	329	335	0
Ag	1036	338	255	297	302	293
AgBr	408	90	85	87	88	293
AgCl	442	86	78	82	82	293
Al	881	287	285	286	286	0
Al	764	263	260	261	261	298
AlSb	593	339	311	325	326	298
Au	1803	336	261	299	305,5	0
$Ba(NO_3)_2$	326	156	145	151	151	293
Diamond	4420	5357	5310	5333	5333	300
$(CH_2)_6N_4$	70	66	66	66	66	293
CaF_2	831	439	406	422	422	293
CaF_2	900	424	400	412	412	300
CaF_2	923	425	404	415	415	220
CaF_2	930	447	421	434	434	100
CaF_2	953	451	426	439	438	4
CsBr	180	106	106	106	106	0
CsI	124	74	71	73	73	295
Cr_2FeO_4	2033	1058	1040	1049	1050	293
Cu	1371	546	400	473	482	293
Cu	1420	593	436	514	523	0
Cu_3Au	1506	556	440	498	505	4
Cu_3Au	1518	511	402	457	464	300
CuZn	1162	533	206	370	386	293

Table I continued

Material	K*	G_V	G_R	G_{VRH}	G*	Temperature(°K)
Iron	1731	941	797	869	875	293
Fe_3O_4	1617	916	911	914	914	293
GaAs	755	486	446	466	467	298
GaSb	564	356	328	342	343	298
Ge	768	574	540	557	558	73
Ge	754	565	531	548	549	273
Ge	739	553	519	536	537	473
Ge	732	546	513	530	530	573
InSb	469	242	217	230	231	298
K	40	17	9	13	14	83
$KAl(SO_4)_2$	156	80	79	80	80	293
KBr	155	89	69	79	77	295
KCl	174	105	84	95	93	293
KCl	197	125	91	108	105	4
KCl	182	106	85	96	94	295
KF	319	179	160	169	169	295
KI	127	85	53	69	66	4
KI	124	69	52	60	60	295
Li	133	69	25	47	49	78
Li	125	64	24	44	46	155
Li	120	62	23	42	44	195
LiBr	257	156	142	149	149	295
LiCl	315	203	185	194	195	295
LiF	698	512	462	487	488	295
LiF	698	554	527	540	541	0
LiI	188	110	100	105	106	295
MgO	1533	1286	1239	1262	1263	293
MgO	1660	1346	1283	1315	1315	293

Table I continued

Material	K*	G_V	G_R	G_{VRH}	G*	Temperature(°K)
MgO	1543	1286	1207	1246	1247	573
MgO	1387	1198	1095	1147	1148	973
Mo	2707	1228	1209	1218	1219	293
Na	66	28	12	20	21	293
NaBr	211	117	113	115	115	295
NaBr	194	116	112	114	114	293
NaCl	252	149	145	147	147	295
NaCl	258	175	161	168	168	0
NaCl	246	150	146	148	148	293
NaF	486	314	308	311	311	295
NaI	161	86	83	84	84	295
NaI	177	108	98	103	103	4
$NaBrO_3$	308	171	168	169	169	293
$NaClO_3$	256	140	135	138	137	293
Ni	1875	1011	848	929	936	4
Nb	1533	457	411	434	437	293
NH_4Br	138	79	68	74	73	293
NH_4Cl	178	104	88	96	95	293
Pd	1954	543	450	497	503	0
Pd	1956	533	437	485	492	80
Pd	1919	529	422	476	484	200
Pd	1923	532	417	475	483	280
Pb	417	101	67	84	87	293
$Pb(NO_3)_2$	370	111	101	106	107	293
PbS	621	343	308	326	325	293
RbBr	138	76	53	65	63	293
RbBr	138	78	55	66	64	295
RbCl	162	90	66	78	76	295

Table I continued

Material	K*	G_V	G_R	G_{VRH}	G*	Temperature(°K)
RbF	273	144	119	131	130	295
RbI	112	59	40	50	48	293
RbI	112	61	40	50	48	295
Si	978	681	650	665	666	293
Ta	1943	711	677	694	696	293
TlBr	225	92	88	90	90	293
TlCl	236	95	90	93	93	293
Th	581	367	259	313	317	0
Th	577	340	233	286	291	293
W	3081	1530	1530	1530	1530	290
V	1543	487	480	483	483	293
ZnS	765	275	248	261	263	293

Table II

all moduli in kilobars

Material	K_V	K_R	K_{VRH}	K*	G_V	G_R	G_{VRH}	G*	Crystal Class	Temperature(°K)
$BaTiO_3$	1073	1071	1072	1072	511	506	509	509	hexagonal	293
Be	1156	1154	1155	1155	1531	1517	1524	1524	hexagonal	0
Be	1155	1153	1154	1154	1530	1516	1523	1523	hexagonal	100
Be	1159	1157	1158	1158	1518	1504	1511	1511	hexagonal	200
Be	1145	1143	1144	1144	1493	1480	1486	1486	hexagonal	300
$CaBaTiO_3$	1106	1106	1106	1106	470	469	469	469	hexagonal	293
Cd	629	532	580	592	319	270	294	295	hexagonal	0
Cd	622	523	572	584	311	262	286	287	hexagonal	80
Cd	603	505	554	567	288	241	264	265	hexagonal	200
Cd	581	488	535	547	264	221	242	243	hexagonal	300
CdS	614	614	614	614	170	166	168	168	hexagonal	293
Co	1905	1904	1904	1904	844	801	822	819	hexagonal	298
Ice	81	81	81	81	37	36	37	37	hexagonal	257
Ice	80	80	80	80	37	36	37	37	hexagonal	248
Ice	78	78	78	78	36	34	35	35	hexagonal	263
Ice	77	77	77	77	35	34	34	35	hexagonal	268

Table II continued

Material	K_V	K_R	K_{VRH}	K*	G_V	G_R	G_{VRH}	G*	Crystal class	temperature(°K)
Mg	369	369	369	369	194	193	193	193	hexagonal	0
SiO_2	566	566	566	566	416	407	412	411	hexagonal	873
SiO_2	565	564	564	564	419	408	413	413	hexagonal	873
Zn	804	661	732	761	511	394	453	467	hexagonal	4
Zn	795	652	723	752	502	383	442	457	hexagonal	77
Zn	775	633	704	733	474	359	416	429	hexagonal	200
Zn	751	616	683	710	448	341	395	406	hexagonal	295
Zn	723	598	661	686	419	321	370	380	hexagonal	400
Zn	696	581	639	662	386	293	340	348	hexagonal	500
Zn	677	566	622	645	351	266	308	316	hexagonal	600
ZnO	1436	1435	1436	1436	458	453		455	hexagonal	293
Y	423	421	422	422	282	280	281	281	hexagonal	4
Y	418	416	417	417	280	278	279	279	hexagonal	75
Y	410	408	409	409	268	266	267	267	hexagonal	200
Y	415	414	415	415	255	253	254	254	hexagonal	300
Y	411	410	411	411	247	245	246	246	hexagonal	400
Graphite	2863	358	1610	1303	2179	7	1093	767	hexagonal	room temp.

Table II continued

Material	K_V	K_R	K_{VRH}	K*	G_V	G_R	G_{VRH}	G*	Crystal Class	Temperature(°K)
$BaTiO_3$	1873	1680	1776	1806	730	621	676	679	tetragonal	298
$BaTiO_3$	1863	1628	1746	1791	599	478	538	538	tetragonal	298
In	464	461	462	463	86	60	73	75	tetragonal	0
In	416	416	416	416	59	37	48	49	tetragonal	298
KH_2PO_4	268	267	267	267	182	128	155	152	tetragonal	293
KH_2PO_4	403	401	402	402	169	123	149	151	tetragonal	280
$NiSO_4 6H_2O$	248	246	247	247	101	76	89	89	tetragonal	293
Sn	438	434	436	436	237	235	236	236	tetragonal	293
Sn	527	526	527	527	191	150	170	172	tetragonal	293
Sn	550	535	543	543	442	410	426	428	tetragonal	293
Sn	579	579	579	579	259	228	244	247	tetragonal	4
TiO_2	2198	2106	2152	2151	1259	1012	1135	1157	tetragonal	293
TiO_2	2156	2102	2129	2128	1173	992	1083	1102	tetragonal	293
Zircon	210	191	200	198	217	184	200	198	tetragonal	293
$(NH_4)H_2PO_4$	290	276	283	284	129	94	112	106	tetragonal	293
$(NH_4)H_2PO_4$	276	265	270	271	119	91	105	101	tetragonal	293
$(NH_4)H_2PO_4$	222	216	219	220	136	97	117	109	tetragonal	293
$(NH_4)H_2PO_4$	263	253	258	259	137	94	115	107	tetragonal	293

Table II continued

Material	K_V	K_R	K_{VRH}	K*	G_V	G_R	G_{VRH}	G*	Crystal Class	Temp. (°K)
$BaSO_4$	530	529	530	530	241	205	223	225	orthorhombic	288
BaSO	589	588	589	589	246	211	228	231	orthorhombic	293
Aragonite	499	455	477	472	412	373	392	390	orthorhombic	293
HiO	207	201	204	204	181	169	175	176	orthorhombic	298
KP borate	274	241	257	260	91	62	76	75	orthorhombic	298
Rochelle salt	379	351	365	366	92	69	81	84	orthorhombic	293
Rochelle salt	395	375	385	386	89	68	79	82	orthorhombic	303
LiA tartrate	160	151	155	155	144	120	132	132	orthorhombic	298
MgSO 7H_2O	440	430	435	436	185	159	172	173	orthorhombic	293
Mg SiO	1337	1289	1313	1312	824	793	809	807	orthorhombic	293
NaA tartrate	365	339	352	353	78	63	71	73	orthorhombic	293
Na tartrate	399	382	390	390	98	74	86	90	orthorhombic	293
S	206	176	191	188	72	62	67	66	orthorhombic	293
ZnSO 7H O	220	213	217	217	140	109	125	128	orthorhombic	293
U	1147	1114	1131	1128	881	807	844	841	orthorhombic	298
SrSO	821	820	820	820	181	86	133	149	orthorhombic	293
Sr dihydrate	116	108	112	112	168	149	158	157	orthorhombic	298
Topaz	1731	1701	1716	1716	1189	1164	1176	1177	orthorhombic	293

Table II continued

Material	K_V	K_R	K_{VRH}	K*	G_V	G_R	G_{VRH}	G*	Crystal Class	Temperature(°K)
Al_2O_3	2514	2509	2511	2511	1660	1607	1634	1634	trigonal	293
Al_2O_3	2504	2499	2501	2501	1626	1574	1600	1601	trigonal	293
$AlPO_4$	755	671	713	721	285	222	254	251	trigonal	293
Bi	347	325	336	338	145	108	126	125	trigonal	300
Bi	360	337	348	347	133	115	122	124	trigonal	293
BrNa dextrose	119	117	118	118	70	69	69	69	trigonal	293
Calcite	692	649	670	674	377	284	331	326	trigonal	293
Calcite	754	713	734	740	358	263	310	304	trigonal	293
ClNa dextrose	126	122	124	124	66	64	65	65	trigonal	293
Fe_2O_3	984	975	980	979	944	914	929	929	trigonal	293
INa dextrose	136	127	132	131	73	67	70	70	trigonal	293
$NaNO_3$	342	294	318	321	264	220	242	241	trigonal	293
SiO_2	380	375	377	377	478	411	444	441	trigonal	293
SiO_2	368	362	365	365	485	416	451	447	trigonal	293
SiO_2	379	374	377	377	479	410	445	441	trigonal	293
Sb	395	362	378	383	251	205	228	227	trigonal	293
Tourmaline	991	886	939	932	895	835	865	865	trigonal	293
Tourmaline	1032	977	1004	1005	891	820	856	855	trigonal	293

REFERENCES

Chapter 2

All textbooks and monographs on probability theory and statistics, for example

A.N. Kolmogorov, Foundations of the Theory of Probability, Chelsea Publishing Company, New York 1956

Yu.V. Prohorov and Yu.A. Rozanov, Probability Theory, Springer-Verlag, Berlin - Heidelberg - New York 1969

D. Morgenstern, Einführung in die Wahrscheinlichkeitsrechnung und mathematische Statistik, Springer - Verlag, Berlin - Heidelberg - New York 1968

H. Cramér, Mathematical Methods of Statistics, Princeton University Press, Princeton, N.J. 1946

M. Loève, Probability Theory, Van Nostrand, Princeton, N.J. 1962

B.W. Lindgren and G.W. McElrath, Introduction to Probability and Statistics, Macmillan Company, New York 1959 (elementary textbook)

The systematics are taken over from

M.J. Beran, Statistical Continuum Theories, Interscience Publishers, New York 1968

Functional analysis is treated by

V. Volterra, Theory of Functionals and of Integral and Integro-Differential Equations, Dover, New York 1959

G. Evans, Functionals and Their Applications, Dover, New York 1964

Chapter 3

M.J. Beran, Statistical Continuum Theories, Interscience Publishers, New York 1968

Chapter 4

Surveys:

G. Batchelor, The Theory of Homogeneous Turbulence, Cambridge University Press, London -New York 1953

M.J. Beran, Statistical Continuum Theories, Interscience Publishers, New York 1968

H. Goering, Sammelband zur Statistischen Theorie der Turbulenz, Berlin 1958

Original work:

G.I. Taylor, Proc. London Math. Soc. 20, 196 (1921)

A.A. Friedmann and L.W. Keller, Proc. Int. Congr. Appl. Mech. Delft 1924

G.I. Taylor, Proc. Roy. Soc. (London) A 151, 421 (1935) and A 156, 307 (1936)

T. von Karman, Proc. Nat. Acad. Sci. U.S. 23, 98 (1937) and J. Aeron. Sci. 4, 131 (1937)

T. von Karman and L. Howarth, Proc. Roy. Soc. (London) A 164, 192 (1938)

A.A. Kolmogorov, Compt. Rend. Acad. Sci. URSS 30, 301 (1941); Engl. Transl. in "Turbulence", S. Friedlander and L. Topper Eds., Interscience Publishers, New York

C.F. von Weizsäcker, Z. Physik 124, 614 (1948)

W. Heisenberg, Z. Physik 124, 628 (1948) and Proc. Roy. Soc. (London) A 195, (1948/49)

E. Hopf, J. Ratl. Mech. Anal. 1, 87 (1952) and Proc. Symp. Appl. Math. 13, 165 (1962), Am. Math. Soc., Providence, R.I.. These papers refer to the functional formulation of the theory of turbulence.

More recent formulations:

H. Wyld, Ann. of Phys. 14, 143 (1961)

V. Tatarski, Wave Propagation in a Turbulent Medium, McGraw Hill, New York 1961

S. Edwards, J. Fluid Mech. 18, 239 (1964)

R. Deissler, Phys. Fluids 8, 291, 2106 (1965)

R. Kraichnan, Phys. Fluids 1, 358 (1958), 7, 1030, 1163, 1723 (1964), 8, 575, 995 (1965), 9, 1728 (1966)

L.S.G. Kovasznay, Turbulence Measurement in Applied Mechanics Surveys, H.N. Abramson. H. Liebowitz, J.M. Crowley, S. Juhasz Eds., Spartan Books, Macmillan and Co. Ltd., Washington, D.C. 1966 (describes experimental technics).

Chapter 5

Surveys:

M.J. Beran, Statistical Continuum Theories, Interscience Publishers, New York 1968.

V.A. Lomakin, Statistical Problems of the Mechanics of Deformable Solids, Acad. Sciences, Moscow 1970 (in Russian).

Original Work on the Statistical theory of Heterogeneous materials:

I.M. Lifshitz and L.N. Rosentsveig, Zh. Eksp. Teor. Phys. 16, 967 (1946).

M. Beran and J. Molyneux, Nuovo Cimento 30, 1406 (1963)

M. Beran, Nuovo Cimento Suppl. 3, 448 (1965)

M. Beran and J. Molyneux, Quart. Appl. Math. 24, 107 (1966)

V.A. Lomakin, Sov. Phys. Dokl. 9, 282 (1964), J. Appl. Math. Mech. (Transl. from Russian) 29, 1048 (1965), Mechanica Polymerov (Riga) 2, 213 (1967)

S.D. Volkov and N.A. Klinskikh, Physics Metals, Metallogr. (Transl. from Russian) 19, 24 (1965)

E. Kröner, J. Mech. Phys. Solids 15, 319 (1967)

K. - H. Reichstein, Zur Statistischen Theorie der Effektiven Dielektrizitätskonstanten Vielkristalliner Stoffe. Dissertation Clausthal. Germany. 1968

M.J. Beran and J.J. McCoy, Int. J. Solids Structures 6, 1035, 1267 (1970)

V.V. Bolotin and V.N. Moskalenko, Sov. Phys. Dokl. 13, 73 (1968) Mech. Twed. Tela 106 (1969).

W.M. Lewin, Prikl. Mat. Mech. 35, 744 (1971).

P. Mazilu, Rev. Roum. Math. Pur. Appl. 17, 261 (1972).

P.H. Dederichs and R. Zeller, KFA-Jül-Report, Jül-877-FF (1972) available at Central Library of KFA Jülich, Germany or from the authors (in German).

R. Zeller and P.H. Dederichs, phys. stat. sol. (b) 1973, in press.

Non-statistical work on heterogeneous materials:

W. Voigt, Abh. Gött. Akad. Wiss. 1887, p. 48

A. Reuss, Z.Angew Math. Mech. 9, 49 (1929)

J.F.W. Bishop and R. Hill, Phil. Mag. 42, 414, 1298 (1951)

R. Hill, Proc. Phys. Soc. A 65, 349 (1952)

G. Bradfield and H. Pursey, Phil. Mag. 44, 437 (1953)

A.V. Hershey, J. Appl. Mech. 21, 236 (1954)

E. Kröner, Z. Phys. 151, 504 (1958)

Z. Hashin and S. Shtrikman, J. Mech. Phys. Solids 10, 343 (1962)

Z. Hashin, Appl. Mech. Rev. 17, 1 (1964) (Review)

R. Hill, J. Mech. Phys. Solids 13, 213 (1965)

B. Budiansky, J. Mech. Phys. Solids 13, 223 (1965)

G. Kneer, Phys. Stat. Sol. 9, 825 (1965)

O.L. Anderson, Physical Acoustics, Vol. III, Part B, W.P. Mason Ed., Academic Press, New York - London 1965

P.R. Morris, Int. J. Eng. Sciences 8, 49 (1970)

H.H. Wawra, Z. Metallkunde 1970

L. Thomsen, J. Geophys. Res., 77, 315 (1972)

Chapter 6

Surveys on Dislocations and Physical Plasticity:

A. Seeger, Encycl. of Physics VII/1 and VII/2, Springer - Verlag Berlin - Göttingen - Heidelberg 1955/58

J. Friedel, Dislocations, Pergamon Press, Oxford etc. 1964

A.H. Cottrell, THeory of Crystal Dislocations, Blackie and Son Ltd., London - Glasgow 1964

J.P. Hirth and P. Lothe, Theory of Dislocations, McGraw Hill, New York 1968

J.J. Gilman, Micromechanics of flow in solids, McGraw Hill, New York 1969

P. Haasen, Dislocations, Physical Chemistry 10, 69 (1970)

Original work (non-statistical)

D.C. Drucker, Structural Mechanics, Pergamon Press, Oxford etc. 1960

E. Kröner, Acta Met. 9, 155 (1961)

B. Budiansky and T.T. Wu, Proc. 4th Congr. Appl. Mech. 1962, p. 1175

J. Zarka, Extrait du Mémorial de l'Artillerie Française, Sciences et Techniques de l'armement 2, 223 (1969)

J.W. Hutchinson, Proc. Roy. Soc. (London) A. 319, 247 (1970)

E. Kröner and C. Teodosiu, Proc. Symp. Found. of Plasticity, Warsaw 1972, Nordhoff, Leyden 1973, in press

Original Work (statistical):

E. Kröner, Initial Studies of a Plasticity Theory Based upon Statistical Mechanics in Inelastic Behaviour of Solids, M.F. Kanninen, W.F. Adler, A.R. Rosenfield, R.I. Jaffee Eds., McGraw Hill Comp., New York 1970.

E. Kröner, Proc. 3rd Disc. Conf. on General Principles of Rheology, Prague 1972, Rheol. Acta 1973, in press.

R. Labusch, Phys. Stat. Sol. 41, 659 (1970).

PROBLEMS

Chapter 2

Section 3

A fundamental problem of probability theory consists in calculating the probabilities of random events from the given probabilities of the elementary events. Answer the following questions;

1) What is the probability that the outcome in the tossing of two dice is

a) a 5 (7, 9) ?

b) an even (odd) number ?

c) 5 or 7 ?

d) not 6 ?

e) larger than 7 ?

2) a) What is the probability that at least two out of N persons have a common birthday ?

(A case is counted as positive also if more than 2 persons have their birthday on the same day or if there exist several days which are common birthdays for 2 or more persons.)

b) What is the probability that exactly two out of N persons have a common birthday ?

Section 4

1) What is the probability that the outcome in the tossing of two dice is 8 if it is already known that the outcome is
   a) an even (odd) number ?
   b) at least 6 ?
   c) divisible by 4.
2) How are the probabilities of problem 2 of Section 3 changed if we have the additional information that not more than 2 persons have their birthday on the same day ?
3) An atom is located at point $x = 0$ at time $t = 0$. In time intervals Δt it jumps with equal probability a distance Δx in the direction of the positive or negative x-axis. What is the probability that it is found at $x = 0$ after 100 jumps?

Section 5

1) Derive the distribution function of problem 3 of Section 4.
2) Derive the distribution function and density function for Gaussian processes.
3) Prove eg. (2.5.11).

Section 6

1) Calculate the expectation and standard deviation for the tossing of two dice.
2) Let the body size in a group of 500 people be described by a

Gaussian distribution with mean 1,75 m and standard deviation 0,1 m. How many people would bump their heads when going upright through a door with height of 1,85 m ? (*)

3) Calculate the characteristic functions belonging to the Gaussian distribution and find the moments $\overline{u^n}$ up to order $n=4$.

4) Calculate the correlation function $\overline{u(x)u(x+\xi)} = f(\xi)$ where $-\infty \leq x \leq +\infty$ and $u(x) = \sin kx$.

What does the result mean ?

5) Give a number of functions $u(x)$ for which the correlation functions $f(\xi)$ can be calculated.

Section 7

1) in the literature on ergodic systems it is shown that the ergodic hypothesis does not apply if not all states are accessible to all members of the ensemble. Give an example for such a situation in the case of a gas in a container.

Section 8

1) A damped harmonic oscillator obeys the differential equation

$$m d^2x/dt^2 + r\, dx/dt + kx = F(t) .$$

Let $F(t)$ be a random driving force with a given spectral density $S_F(\nu)$ at frequency ν .

a) Calculate the spectral density $S_x(\nu)$ of the displacement x.

b) Calculate the standard deviation of $x(t)$ in the special

(*) After B.W. Lindgnen and G.W. Elrath

case of a "white" force spectrum of $F(t)$ i.e. if $S_F(\nu) =$
$=$ constant.

c) Discuss the physical implications of the solution obtained.

Section 9

1) Establish the connection between the probability density functional $P[u(t)]$ and the set of correlation functions of $u(t)$ by means of a Taylor development of $P[u(t)]$.

Chapter 3

Section 12

1) Derive the equation for the 2-point correlation function from eq. (3.12.1) and form the Fourier transform of this equation under the assumption of a spatially stationary situation. Try to find a physical meaning for your results.

Chapter 4

Section 15

1) Prove eq. (4.15.17)
2) Prove eq. (4.15.28) and (4.15.29) and compare your consideration with that used by Beran in his derivation of those equations.

Chapter 5

Section 16

1) In his theory of the Voigt and Reuss bounds, Hill assumes the validity of eq. (5.16.7). He shows that

$$<\underline{\varepsilon}> \underline{C} <\underline{\varepsilon}> \equiv <\underline{\varepsilon}\, \underline{C}\, \underline{\varepsilon}> \leqq <\underline{\varepsilon}><\underline{C}><\underline{\varepsilon}>$$

$$<\underline{\sigma}> \underline{C}^{-1} <\underline{\sigma}> \equiv <\underline{\sigma}\, \underline{C}^{-1} \underline{\sigma}> \leqq <\underline{\sigma}><\underline{C}^{-1}><\underline{\sigma}> .$$

Prove that this is right.

Section 18

1) Approximate values of the effective elastic moduli can be obtained by taking the geometric mean of the Voigt and Reuss bounds. In which power of the anisotropy constants do these means deviate from the true effective moduli of perfectly disordered polycristalline aggregates ?
2) Since the local compression modulus of a single cubic crystal and the effective compression modulus of an aggregate of such crystals coincide, only one equation is needed to relate the effective shear modulus G with the single crystal moduli. Recently, Peresada (Phys. Stat. Sol. 1971) has proposed to use the equation obtained by equating the determinants of the local and the effective tensor of the elastic moduli. Examine if this is a good approximation for small anisotropy.

3) Does perfect disorder imply macroscopic isotropy?

Chapter 6

Section 21

1) Show that $< t_i b_j t_i b_j >_{\underline{x}}$ is proportional to the total dislocation length per unit volume (cf. E. Kröner 1970, l.c.)

CONTENTS

Printe
By B